地理科学类专业实验教学丛书

GIS 专题开发与设计实验教程

胡引翠 刘 强 等 编著

科学出版社
北 京

内 容 简 介

本书通过具体实验教学案例，介绍桌面 GIS、WebGIS、移动 GIS 等地理信息系统开发的技术方法和实现步骤。本书第一部分主要介绍如何利用开源组件 DotSpatial 创建桌面 GIS 开发，第二部分介绍基于 ArcGIS Engine 的 WebGIS 开发，第三部分介绍基于 GeoServer 的 WebGIS 开发，第四部分介绍基于 ArcGIS 的移动 GIS 开发。

本书可作为高等院校地理科学、地理信息科学及相关专业本科生的实验教材，也可作为地理信息系统开发技术人员的培训教材和参考书。

图书在版编目（CIP）数据

GIS 专题开发与设计实验教程/胡引翠等编著. —北京：科学出版社，2021.10

（地理科学类专业实验教学丛书）

ISBN 978-7-03-070041-4

Ⅰ.①G… Ⅱ.①胡… Ⅲ.①地理信息系统-系统开发-应用软件-高等学校-教材 Ⅳ.①P208

中国版本图书馆 CIP 数据核字（2021）第 211231 号

责任编辑：杨 红 郑欣虹/责任校对：杨 赛
责任印制：张 伟/封面设计：陈 敬

科学出版社 出版
北京东黄城根北街 16 号
邮政编码：100717
http://www.sciencep.com

北京九州迅驰传媒文化有限公司 印刷
科学出版社发行 各地新华书店经销

*

2021 年 10 月第 一 版　开本：787×1092　1/16
2022 年 1 月第二次印刷　印张：13 1/2
字数：324 000

定价：59.00 元
（如有印装质量问题，我社负责调换）

河北省地理科学实验教学中心建设成果
河北省环境变化遥感识别技术创新中心建设成果
"信息地理"河北省优秀教学团队建设成果

"地理科学类专业实验教学丛书"编写委员会

主　　编：李仁杰

副主编：张军海　任国荣

编　　委：（按姓名汉语拼音排序）

常春平　丁疆辉　傅学庆　郭中领　韩　倩
胡引翠　李继峰　李仁杰　刘　强　任国荣
田　冰　王文刚　王锡平　严正峰　袁金国
张鉴达　张军海　郑东博

丛 书 前 言

在移动互联网飞速发展的今天，学生可以获取的教学资源日益丰富，教学模式趋向多元化，慕课、微课、共享资源课、VR 教学等新型教学方式极大地方便了学生的课堂外自主学习。传统课堂教学面临前所未有的挑战，许多教师也在尝试引入这些新的教学资源和方法，以适应时代的发展。但无论如何，大学教育的一个核心教学思想不会改变，那就是通过教学过程帮助学生建构学科知识体系，培育专业学科素养和创新性思维。教学过程使用的各类教学资源中，教材是支撑这一核心教学思想的最重要资源。无论传统纸质教材与现在流行的电子书形式差别有多大，都必须达到支撑上述思想的标准。

地理学的特点是综合性和区域性，地球表层系统空间各个要素不仅具有自身的空间分布格局与特征，也同其他地理要素具有空间联系并相互影响。地理学科的专业教材不仅要专注于解析地理学某一分支学科的知识体系，更应帮助学生建构与其他分支学科的关系。例如，自然地理学与人文地理学两门课程，既有相对独立的学科思想、理论和方法，也有共同的研究对象，我们可以借助全球变化研究中关于人类活动的环境响应等主题，实现两个分支学科关系的知识体系建构，进而培养学生综合性学术思维。

大学地理科学相关专业的课程实验是从理论到实践的教学过程，通过实验教学帮助学生深入理解其所建构的学科知识体系，完成基于理论方法解决实际学科问题的训练过程，并能够独立解决新问题，这是实验教学资源（特别是实验教程）应该实现的基本功能。

河北师范大学资源与环境科学学院的地理科学相关专业已有 60 多年的办学历史，一批批地理学者以科学严谨的学术探索和言传身教的人才培育为己任，笔耕不辍，出版了不少经典学术著作和优秀的教材。如今，学院继续蓬勃发展，2011 年获得地理学一级学科博士学位授予权，2014 年获批地理学博士后科研流动站，新的一批年轻地理学者也已经成长起来，风华正茂，希望他们能够继承优良传统，成就新的辉煌。恰逢 2015 年学院获批河北省地理科学实验教学示范中心，如何将优秀教学理念与方法向社会传播，实现优质教学资源的共建与分享，成为年轻一代教师们思考的重要问题。从当代地理科学发展的现状来看，大家一致认为，应该着重构建学生实践创新能力培养的多元化实验教学环境，将地理信息科学专业的实验教学作为示范中心重点培育的纽带项目，充分发掘互联网服务资源与功能，整合地理信息科学、自然地理学、人文地理学和其他相关学科的实验教学内容，逐步构建"多专业实验协同创新与环境共享的实验教学体系"，推进"教师科研创新引领下的实验教学改革模式"，全面实现示范中心教学资源共享。

任重而道远，我们必须脚踏实地，砥砺前行。地理科学实验教学系列教材的编著工作正式启动了。系列中的每本实验教程都不是对单一课程的独立实验描述，而是按照学科体系将学科知识关系密切的相关课程集成在一起，统一设计实验项目和内容。每本教材的内容设计与系列教材的总体架构，就是引导学生建构课程知识体系和培养学科思维模式的双层脉络。例如，地图学、空间数据管理与可视化和地理信息系统原理三门课程的集成实验，遥感导论与遥感数字图像处理两门课程的集成实验，测量学、全球导航定位系统原理和数字摄影测量三门课程的集成实验，以及地理信息数据挖掘与软件开发相关的课程集成实验等。

特别需要说明的是，实验教材系列中还有一本有关典型实验数据集的教材，数据来源包括政府开放数据（如社会经济统计数据）、科学共享数据（如全球 30m 分辨率数字地形、地表覆盖数据）、志愿者地理信息数据（雅虎 YFCC 数据集）等，这些典型数据集不仅可以支撑众多相关课程的实验教学训练，还可以帮助学有余力的同学寻找科学问题，开展创新性地理研究探索。

这套系列教材的执笔者都对大学教育情有独钟，他们中既有已过知天命之年阅历丰富的教授，他们不忘初心，继续编写教材令人敬佩；也有肩负行政管理、科学研究和本科教学多重任务的中青年骨干，他们在繁重的工作中不求名利，守望净土，让人欣喜；更有刚刚入职的青年才俊，他们初生牛犊、意气风发，使人振奋。整套系列教材完全编写完毕会超过 20 本的规模，以地理信息科学专业的 9 本实验教材为主，再加上前期积累较好的地理教育教学实践教材，作为引领启动的一期工程。二期工程将以地理学科相关本科专业的核心课程为基础，整合实验室基础实验和野外实习实验，并与一期工程的相关教材形成内容互补、体系呼应的整体成果。希望通过大家的努力，影响更多教师投入到系列教材编写中，为地理科学专业人才培养做出贡献。当然，我们不追求教材的形式，正如开篇所述，无论是纸质书还是电子书，还是直接发布到互联网进行共享和传播教材资源，最重要的是教材要有设计思想，要以合适的形式不断发展演进，主动适应快速变化的学科理论和方法，要能够支持慕课、微课和 VR 教学等各种新型教学模式，最终以培养学生的创新性思维和专业素养为最高价值目标。

李仁杰
2018 年 8 月

前　　言

　　GIS 是伴随着计算机技术和通信技术的发展而演进的。每一次技术变革都为 GIS 带来新的活力和发展方向。计算机技术的发展，使 GIS 从早期依托的大型机、个人计算机发展到现在的移动智能终端设备。通信技术的发展，使 GIS 从单机版、工作站版发展到现在的网络服务。计算机技术、通信技术和位置服务相关技术的发展，使 GIS 从专业化的小众群体走向大众视野。

　　GIS 专题开发与设计实验是地图学、地理信息系统等前期课程基础上的进阶课程。作为 GIS 专业的核心课程之一，GIS 专题开发与设计课程要求学生具备计算机软件开发基础并且熟练掌握 GIS 基础理论，学习难度较大。为了让学生更好更快地具备 GIS 开发能力，河北师范大学资源与环境科学学院地理信息教研室组织多方力量，成立了实验教程编写小组。

　　本书内容涵盖了桌面 GIS 开发、WebGIS 开发和移动 GIS 开发实验，兼具容纳了应用较为广泛的商业软件和开源软件。受能力所限，本书不能尽善尽美，但作者尽力向学生呈现多样化的 GIS 开发途径，期望能够引领热爱 GIS 专业的学生开启开发之门。

　　本书包括四部分内容：①基于 DotSpatial 的桌面 GIS 开发系列实验。本部分实验基于开源组件，主要介绍地理数据加载浏览与地图浏览、地理要素查询与检索、专题地图渲染、地理要素编辑、空间分析与统计、地图打印输出等软件开发内容。②基于 ArcGIS Engine 的开发系列实验。本部分实验基于商业软件组件开发，主要包括 ArcGIS Engine 控件的使用、地图文档及相关对象、几何对象与空间参考、参数传递与鹰眼地图、命令封装与右键菜单、空间可视化、空间数据库、空间数据查询、空间数据编辑、空间分析、地图整饰输出等内容。③基于 GeoServer 的 WebGIS 开发系列实验。本部分实验主要包括 GeoServer 的安装与使用、地图图层的发布与管理、地图浏览、地图图层叠加显示、屏幕交互操作等内容。④基于 ArcGIS 的 Android 移动 GIS 开发系列实验。本部分实验主要包括 ArcGIS SDK for Android 开发环境及配置、地图工程创建、数据显示与浏览、数据查询与检索、数据采集与编辑、数据实时同步等内容。

　　本书由胡引翠设计、统稿和定稿。具体编写任务分工如下：第一部分由胡引翠设计完成，软件代码由郝斌编写实现；第二部分由刘强设计完成；第三部分由胡引翠、刘强共同设计完成；第四部分由胡引翠设计完成。

　　在本书编写过程中，作者课题组的多位研究生帮助完成了相关实验的测试和数据整理及校对工作，包括已毕业的硕士研究生：郝斌、任致华、刘立亚、高博、武紫超、程雅琪、高玉健、刘学、王云、王旭、薛瑞恒等，以及在读硕士研究生王天宇、王晨旭、张文静、鲍艳磊、王博伟、顾敏珺、刘朝阳、张征南、丁梦婷等。特别感谢张金星、杨娜娜、胡士佳、梁亚南、路志娜、赵瑞锋、李玉森、李雷星参与软件测试和代码校对工作，他们耐心、仔细和一丝不苟的态度让作者感动。本书编写过程中还有很多老师和同学提供了建议和帮助，不能一一列出，在此一并表示感谢！

　　本书是编写组根据多年的教学实践，在听取学生、同行的建议下编写完成的。书中涉及的章节实验经多次讨论调整，几经修改完成。希望本书的出版，能够为系统化的 GIS 专业教学设计提供帮助和参考。

本书实验中所涉及的代码和工程文件可通过河北师范大学资源与环境科学学院网站下载，地址：http://zhxy.hebtu.edu.cn/。虽然所有代码均已进行测试，但受作者水平所限，错误与不妥之处在所难免，敬请读者批评指正。对本书有任何建议请发信至 huyincui@163.com。

作　者
2020 年 9 月

目　录

丛书前言
前言

第一部分　基于 DotSpatial 的桌面 GIS 开发系列实验 ……………………………… 1
实验 1-1　地理数据加载与地图浏览 ……………………………… 1
实验 1-2　地理要素查询与检索 ……………………………… 5
实验 1-3　专题地图渲染 ……………………………… 10
实验 1-4　地理要素编辑 ……………………………… 17
实验 1-5　空间分析与统计 ……………………………… 25
实验 1-6　地图打印输出 ……………………………… 30

第二部分　基于 ArcGIS Engine 的开发系列实验 ……………………………… 38
实验 2-1　ArcGIS Engine 控件的使用 ……………………………… 38
实验 2-2　地图文档及相关对象 ……………………………… 47
实验 2-3　几何对象与空间参考 ……………………………… 55
实验 2-4　参数传递与鹰眼地图 ……………………………… 67
实验 2-5　命令封装与右键菜单 ……………………………… 71
实验 2-6　空间可视化 ……………………………… 80
实验 2-7　空间数据库 ……………………………… 91
实验 2-8　空间数据查询 ……………………………… 98
实验 2-9　空间数据编辑 ……………………………… 105
实验 2-10　空间分析 ……………………………… 112
实验 2-11　地图整饰输出 ……………………………… 120

第三部分　基于 GeoServer 的 WebGIS 开发系列实验 ……………………………… 131
实验 3-1　GeoServer 的安装与使用 ……………………………… 131
实验 3-2　地图图层的发布与管理 ……………………………… 134
实验 3-3　地图浏览 ……………………………… 139
实验 3-4　地图图层叠加显示 ……………………………… 146
实验 3-5　屏幕交互操作 ……………………………… 158

第四部分　基于 ArcGIS 的 Android 移动 GIS 开发系列实验 ……………………………… 166
实验 4-1　ArcGIS SDK for Android 开发环境及配置 ……………………………… 166
实验 4-2　地图工程创建 ……………………………… 171
实验 4-3　数据显示与浏览 ……………………………… 174

实验 4-4　数据查询与检索……………………………………………………………190

实验 4-5　数据采集与编辑……………………………………………………………197

实验 4-6　数据实时同步………………………………………………………………199

主要参考文献……………………………………………………………………………203

第一部分 基于 DotSpatial 的桌面 GIS 开发系列实验

实验 1-1 地理数据加载与地图浏览

DotSpatial 以控件方式提供地理数据的加载、展示和分析功能。DotSpatial 是开源软件，组件核心为 DotSpatial.Controls.dll 动态链接库，特点是代码完全开放，类库可供其他程序直接调用。使用者可以从其官网上下载源代码供个人修改研发。

（1）实验目的：通过地理数据加载与地图浏览实习，熟悉在.NET 环境和 Visual Studio 平台下搭建和运行 DotSpatial 类库，初步了解开源桌面 GIS 的二次开发模式，初步认识 DotSpatial 类库的基本组织架构，初步掌握 DotSpatial 主要控件的功能和特性，并使用 DotSpatial 类库加载矢量和栅格数据的方法，实现对地图视图的基本浏览操作。

（2）相关实验：GIS 专业实验设备与环境配置中的"GIS 应用开发环境"和"GIS 应用开发资源"。

（3）实验数据：本教材系列实验数据。

（4）实验环境：Visual Studio2010、DotSpatial 1.7 库、.NET Framework 4.0 框架、C#编程语言。

（5）实验内容：通过在 Visual Studio 下加载 DotSpatial 类库，添加 DotSpatial 控件，初步熟悉利用 DotSpatial 进行二次开发的环境搭建过程和控件的简单操作；通过在 Visual Studio 下使用 Map 控件和代码编写一个窗体应用程序实现加载矢量和栅格数据的功能；通过在 Visual Studio 下使用 Map 控件和代码编写一个窗体应用程序实现对地图的放大、缩小、漫游、居中等基本视图操作功能。

1. 环境初识

启动 Visual Studio2010，依次选择"文件"→"新建"→"项目"，弹出"新建项目"对话框，如图 1-1-1 所示。选择 Visual C#项目模板，再选择项目类型为 Windows 窗体应用程序，

图 1-1-1 新建窗体应用程序界面

输入项目名称并选择项目位置，点击"确定"按钮后，则成功新建一个 WinFrom 窗体程序，并自动生成一个名为 From1 的窗体。

在窗体设计界面的工具箱上右击空白处，选择"添加选项卡"并命名为"DotSpatial"，鼠标右击 DotSpatial 选项卡，选择"选择项"，则弹出"选择工具箱项"对话框。选择".NET Framework 组件"，如图 1-1-2 所示，点击"浏览"按钮，在弹出的对话框中找到本地的 DotSpatial.Controls.dll 文件（其打包了一系列 GIS 桌面开发所需的基本控件），如图 1-1-3 所示，点击"确定"后工具箱中出现常用的 DotSpatial 相关控件。

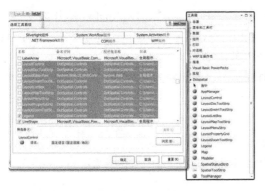 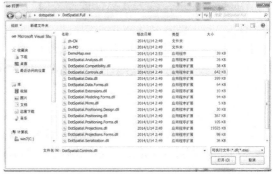

图 1-1-2　工具箱加载 DotSpatial 控件界面　　　图 1-1-3　本地 DotSpatial.Controls.dll 文件界面

右击解决方案资源管理器中的项目名称，选择添加引用，从"浏览"中找到本地的 4 个文件，如图 1-1-4 所示，点击"确定"后，工程引用中出现了添加的 DotSpatial 库。

设置好后，右击解决方案中的项目名称，选择属性，如图 1-1-5 所示，在属性窗体中选择目标框架为".NET Framework 4"版本后，将配置完成 DotSpatial 的开发环境。

图 1-1-4　添加 DotSpatial 库引用界面　　　　　图 1-1-5　项目属性界面

自主练习：创建新工程，配置完成 DotSpatial 的开发环境。

2. 地理数据加载

DotSpatial 可加载矢量和栅格数据：矢量数据支持 ERSI 公司的 shp 格式，而栅格数据支持 bgd、tif、jpg、bmp 等多种格式。DotSpatial 自带 Legend 控件，不需要编写烦琐的代码程序就能实现基本的图层管理功能。程序设计时可拖拽 Legend 控件到主窗体上，并将 Map 控件属性中的 Legend 属性设置为 Legend1 即可。但通常利用代码方式进行程序开发，本实验将全部使用代码编写。

将工具箱中的 Map 控件拖拽到主窗体中，再拖拽两个 Button 按钮到主窗体中，分别命名为"btnLoad"和"btnClear"，按钮文本为"添加图层"和"删除图层"。分别双击"添加图层"和"删除图层"按钮，并在按钮的事件中编写相应的代码。

DotSpatial 提供了如下加载不同类型数据的方法：

map1.AddLayer();	//加载单个矢量或栅格图层
map1.AddLayers();	//加载多个矢量或栅格图层
map1.AddFeatureLayer();	//加载单个矢量图层
map1.AddRasterLayer();	//加载单个栅格图层
map1.AddFeatureLayers();	//加载多个矢量图层
map1.AddRasterLayers();	//加载多个栅格图层
map1.AddImageLayer();	//加载单个位图图层
map1.AddImageLayers();	//加载多个位图图层
map1.ClearLayers();	//删除所有图层

本实验代码如下：

```csharp
using DotSpatial.Controls;
//加载矢量或栅格数据
private void btnLoad_Click(object sender, EventArgs e)
{
    map1.AddLayer();
}
//删除所有图层
private void btnClear_Click(object sender, EventArgs e)
{
    map1.ClearLayers();
}
```

点击 Visual Studio 的启动调试项，运行程序，出现程序主界面，点击"添加图层"按钮，在弹出的选择文件对话框中分别选择本地 shp 和 tif 格式的数据，矢量数据和栅格数据添加分别如图 1-1-6 和图 1-1-7 所示。Map 控件中将分别显示矢量和栅格图层。点击"删除图层"按钮将删除界面的所有图层。

自主练习：完善按钮的事件中对应的添加地图数据的相应代码，实现添加矢量数据、栅格数据。

3. 地图浏览

图层视图操作只需调用 DotSpatial 已经封装好的方法即可，如放大、缩小、漫游、居中、选择等功能。

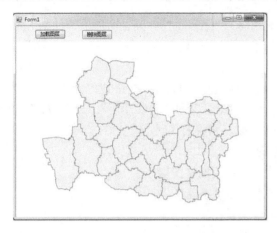 

图 1-1-6　添加矢量数据界面　　　　　图 1-1-7　添加栅格数据界面

拖拽 4 个 Button 按钮到主窗体中，分别命名为"btnZoomIn""btnZoomOut""btnPan" "btnExtent"，按钮文本分别为"放大""缩小""漫游""居中"。分别双击各个按钮，并在按钮的事件中编写相应的代码。

实验代码如下：

```csharp
using DotSpatial.Controls;
//放大地图
private void btnZoomIn_Click(object sender, EventArgs e)
{
    map1.FunctionMode = FunctionMode.ZoomIn;
}
//缩小地图
private void btnZoomOut_Click(object sender, EventArgs e)
{
    map1.FunctionMode = FunctionMode.ZoomOut;
}
//漫游地图
private void btnPan_Click(object sender, EventArgs e)
{
    map1.FunctionMode = FunctionMode.Pan;
}
//居中显示地图
private void btnExtent_Click(object sender, EventArgs e)
{
    map1.ZoomToMaxExtent();
}
```

运行程序，加载地图，点击"放大"按钮，在界面上选中要放大的区域后松开鼠标，该区域将按比例放大；而点击"漫游"按钮，则可拖动鼠标实现地图平移。点击其他按钮执行操作可出现相应结果。放大、居中、漫游等代码运行效果如图 1-1-8～图 1-1-10 所示。

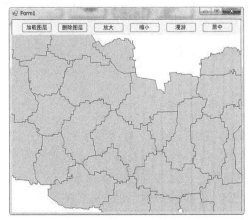
图 1-1-8　放大地图界面

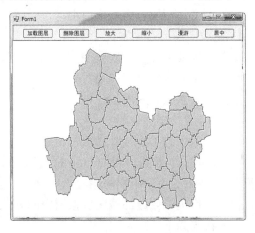
图 1-1-9　居中显示地图界面

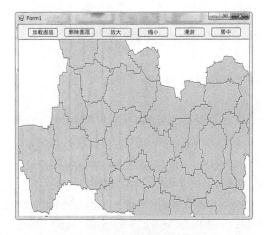
图 1-1-10　漫游地图界面

自主练习：完善各按钮的事件中对应的地图浏览的相应代码，实现数据的放大、缩小、漫游、全图显示。

实验 1-2　地理要素查询与检索

DotSpatial 的 Map 控件中，已集成好图形查询功能，点击地图区域即可弹出属性对话框，但通常需要在属性表窗体上进行属性值修改。本节实验主要完成地理要素查询检索功能的开发。

（1）实验目的：通过地理要素查询与检索实验，进一步了解 DotSpatial 主要控件的功能和特性，初步掌握使用 DotSpatial 类库从空间图形查询属性和从属性查询检索图形信息的方法。

（2）相关实验：实验 1-1 地理数据加载与地图浏览。

（3）实验数据：本教材系列实验数据。

（4）实验环境：Visual Studio2010、DotSpatial 1.7 库、.NET Framework 4.0 框架、C#编程语言。

（5）实验内容：通过在 Visual Studio 下使用 Map 控件和代码编写一个窗体应用程序，实现由点选和框选图形查询空间数据属性的功能；通过在 Visual Studio 下使用 Map 控件和代码编写一个窗体应用程序，实现根据属性条件查询图形和检索空间数据的功能。

1. 由空间图形查询属性

本节实验要求编写代码实现图形选择、弹出属性表窗体功能。

弹框显示要素信息代码为

this.map1.FunctionMode = FunctionMode.Info;

1）选择图斑事件编辑

按照本章实验 1-1 的步骤添加 Map 控件和加载、删除、放大、缩小图层的 Button 按钮并编写好相应代码。拖拽一个 Button 按钮到主窗体中，命名为"btnSelect"，按钮文本为"选择图斑"，在按钮的事件中编写相应的代码。右击主窗体，在属性事件中双击 Form1_Load 编写窗体初始化代码，修改加载图层按钮中的方法并绑定到 Map 控件中的 Load 事件下。

实验代码如下：

```
using DotSpatial.Controls;
//定义当前选择图层
IMapFeatureLayer    currentLayer = null;
//定义选择图斑方法
MapFunctionSelect    selectFunction = null;
//窗体初始化
private void Form1_Load(object sender, EventArgs e)
{
    selectFunction1 = new MapFunctionSelect(this.map1);
    this.map1.MapFunctions.Add(selectFunction);
    this.map1.ProjectionModeReproject = ActionMode.Never;
    this.map1.ProjectionModeDefine = ActionMode.Never;
}
//选择图斑
private void btnSelect_Click(object sender, EventArgs e)
{
    this.map1.FunctionMode = FunctionMode.None;
    selectFunction.Activate();
}
//加载图层后绑定选择图层事件
private void btnLoad_Click(object sender, EventArgs e)
{
    currentLayer = map1.AddLayer() as IMapFeatureLayer;
    selectFunction.Layer = currentLayer;
}
```

2）更改选择图斑的方法

在项目名上右击，添加一个新的 WinFrom 窗体，命名为"Attributes"，从工具箱中拖拽一个 DataGridView 控件到 Attributes 窗体上，设置 DataGridView 控件的"Dock"属性为"Fill"，确保控件始终占满整个窗体。在窗体的 Form1_Load 事件中绑定选择图层的委托事件，在 DataGridView 控件的 SelectionChanged 事件中编写更改选择图斑的方法。

实验代码如下：

```
using DotSpatial.Controls;
using DotSpatial.Data;
//窗体初始化
private void Attributes_Load(object sender, EventArgs e)
{
    layer.SelectionChanged += new EventHandler(currentLayer_SelectionChanged);
}
//设置 DataGridView 数据源
public DataTable WantToDisplayData
{
    set
    {
        this.dgvAttributeTable.DataSource = value;
    }
}
//选择图斑改变方法
public void currentLayer_SelectionChanged(object sender, EventArgs e)
{
    DataTable dt = new DataTable();
    List<IFeature> features = layer.Selection.ToFeatureList();
    dt.Columns.AddRange(layer.DataSet.GetColumns());
    for (int i = 0; i < features.Count; i++)
    {
        DataRow dr = dt.NewRow();
        dr.ItemArray = features[i].DataRow.ItemArray;
        dt.Rows.Add(dr.ItemArray);
    }
    dgvAttributeTable.DataSource = dt;
}
```

3）显示选择图斑属性

在项目名上右击，添加一个类，命名为"MapFunctionSelect"。因为 DotSpatial 内部已集成点选和拉框选择图斑的方法，所以可直接复制进行修改。此方法代码比较复杂，这里只展示修改的代码部分。

实验代码如下：

```
//展示选择图斑的属性
Attributes attributes = null;
if (attributes != null)
{
    attributes.Close();
    attributes = null;
}
attributes = new Attributes();
DataTable dt = new DataTable();
List<IFeature> features = layer.Selection.ToFeatureList();
dt.Columns.AddRange(_layer.DataSet.GetColumns());
for (int i = 0; i < features.Count; i++)
{
    DataRow dr = dt.NewRow();
    dr.ItemArray = features[i].DataRow.ItemArray;
    dt.Rows.Add(dr.ItemArray);
}
attributes.WantToDisplayData = dt;
attributes.Show();
```

运行程序，加载一个地图，点击"选择图斑"按钮，在地图上点击一个图斑后将弹出这个图斑的属性表，再次点击其他图斑，属性表中的内容将替换为新选择图斑的属性；在地图上拉框选择，将弹出包含所选区域的图斑属性表，再次拉框选择则属性表内容替换为新区域的图斑属性表。点选和框选图斑结果及属性分别如图1-2-1和图1-2-2所示。

图1-2-1　点选结果及属性界面

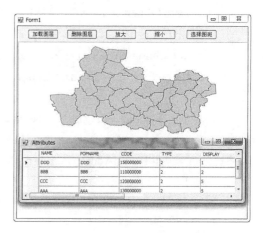

图1-2-2　框选结果及属性界面

自主练习：对照编程实现由空间图形查询属性的地理信息系统功能。

2. 由属性查询检索图形信息

调用 DotSpatial 库的 FilterExpression 方法，可以实现根据属性条件查询空间图形的功能。本实验实现基于单值条件和范围区间条件检索图形信息的功能。

拖拽两个 Button 按钮到主窗体中，分别命名为"btnUniquefind"和"btnRangefind"，按钮文本为"唯一值查询"和"区间查询"，在按钮的事件中编写相应的代码。

实验代码如下：

```csharp
using DotSpatial.Controls;
using DotSpatial.Symbology;
//唯一值查询
private void btnUniquefind_Click(object sender, EventArgs e)
{
    if (map1.Layers.Count > 0)
    {
        MapPolygonLayer mapLayer = default(MapPolygonLayer);
        mapLayer = (MapPolygonLayer)map1.Layers[0];
        mapLayer.SelectByAttribute("[NAME] = '第一分区'");
    }
    else
    {
        MessageBox.Show("图层不存在！");
    }
}
//区间查询
private void btnRangefind_Click(object sender, EventArgs e)
{
    if (map1.Layers.Count > 0)
    {
        MapPolygonLayer mapLayer = default(MapPolygonLayer);
        mapLayer = (MapPolygonLayer)map1.Layers[0];
        mapLayer.DataSet.FillAttributes();
        PolygonScheme scheme = new PolygonScheme();
        PolygonCategory category = new PolygonCategory(Color.Yellow, Color.Red, 1);
        string filter = "[CODE] < 1000 ";
        category.FilterExpression = filter;
        category.LegendText = "population > 10 Million";
        scheme.AddCategory(category);
        mapLayer.Symbology = scheme;
    }
    else
```

```
            {
                MessageBox.Show("图层不存在！");
            }
        }
```

运行程序，加载地图，点击"唯一值查询"按钮，地图上满足"[NAME] = '第一分区'"查询条件的"第一分区"区域将高亮显示，如图 1-2-3 所示；点击"区间查询"按钮，地图上满足"[CODE] < 1000 "查询条件的区域将高亮显示，如图 1-2-4 所示。

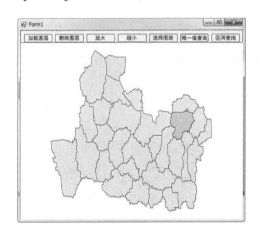 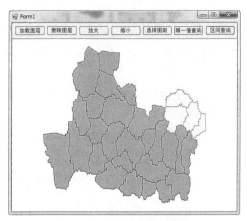

图 1-2-3　唯一值查询界面　　　　　　　图 1-2-4　区间查询界面

自主练习：对照编程实现由属性查询检索图形信息，并定位到所选图形的地理信息系统功能。

实验 1-3　专题地图渲染

DotSpatial 的 Symbology 类库中提供了颜色、字符、图片、梯度等多种不同模式的渲染方式，并可以对点、线、面三种矢量图层进行渲染。

（1）实验目的：通过专题地图渲染实习，进一步了解 DotSpatial 主要控件的功能和特性，初步了解地图符号化和专题地图渲染的原理，初步掌握使用 DotSpatial 类库由颜色和自定义符号两种方式渲染专题地图的方法。

（2）相关实验：实验 1-1 地理数据加载与地图浏览、实验 1-2 地理要素查询与检索。

（3）实验数据：本教材系列实验数据。

（4）实验环境：Visual Studio2010、DotSpatial 1.7 库、.NET Framework 4.0 框架、C#编程语言。

（5）实验内容：通过在 Visual Studio 下使用 Map 控件和代码编写一个窗体应用程序，实现根据不同颜色对矢量地图进行渲染的功能；通过在 Visual Studio 下使用 Map 控件、Legend 控件和代码编写一个窗体应用程序，实现根据自定义符号对矢量图层进行渲染的功能。

1. 矢量地图渲染

在 Symbology.Forms 类库中已集成点、线、面等地图渲染组件，可通过二次开发实现自定义方式。这里通过随机颜色来实现矢量地图渲染的功能。

拖拽一个 Button 按钮到主窗体中，命名为"btnColorRenderer"，按钮文本为"颜色渲染"，在按钮的事件中编写相应的代码。

实验代码如下：

```csharp
using DotSpatial.Controls;
using DotSpatial.Symbology;
//按随机颜色渲染地图
private void btnColorRenderer_Click(object sender, EventArgs e)
{
    if (map1.Layers.Count > 0)
    {
        MapPolygonLayer mapLayer = default(MapPolygonLayer);
        mapLayer = (MapPolygonLayer)map1.Layers[0];
        PolygonScheme scheme = new PolygonScheme();
        scheme.EditorSettings.ClassificationType = ClassificationType.UniqueValues;
        scheme.EditorSettings.FieldName = "NAME";
        scheme.CreateCategories(mapLayer.DataSet.DataTable);
        mapLayer.Symbology = scheme;
    }
    else
    {
        MessageBox.Show("图层不存在！");
    }
}
```

运行程序，加载地图，点击"颜色渲染"按钮，矢量地图的不同图斑依随机颜色渲染，运行效果如图 1-3-1 所示。

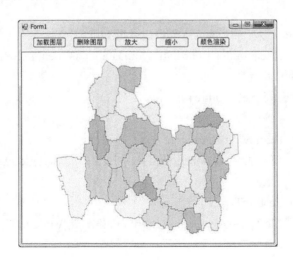

图 1-3-1　随机颜色渲染地图界面

自主练习：对照编程，对加载的矢量数据实现专题渲染。

2. 自定义符号渲染

DotSpatial 没有提供直接加载符号的格式文件，但通过 Serialization 类库中的序列化操作将符号的参数存储到 XML 文件中可自定义符号格式。结合 DotSpatial 的 Legend 控件中的图例对象可修改符号并保存为新的符号文件。

将工具箱中的 Legend 控件拖拽到主窗体中，在 Map 控件的 Legend 中添加 legend1 属性；再拖拽两个 Button 按钮到主窗体中，分别命名为"btnStyleRenderer"和"btnSaveStyle"，按钮文本为"符号渲染"和"保存符号"，在按钮的事件中编写相应的代码。

实验代码如下：

```
using DotSpatial.Controls;
using DotSpatial.Symbology;
using DotSpatial.Serialization;
using DotSpatial.Data;
using DotSpatial.Topology;
//加载地图按钮，添加渲染属性筛选器
private void btnLoad_Click(object sender, EventArgs e)
{
    currentLayer = map1.AddLayer() as IMapFeatureLayer;
    //设置属性筛选器
    currentLayer.Symbology.EditorSettings.ClassificationType=
    ClassificationType.UniqueValues;
}

//自定义符号唯一值渲染地图
private void btnStyleRenderer_Click(object sender, EventArgs e)
{
    XmlDeserializer xd = new XmlDeserializer();
    IFeatureSet featureSet = currentLayer.DataSet as IFeatureSet;
    string filePath = "";
    switch (featureSet.FeatureType)
    {
        case FeatureType.Polygon:
            filePath = Application.StartupPath + @"\style\td09.style";
            break;
        case FeatureType.Line:
            filePath = Application.StartupPath + @"\style\td07.style";
            break;
        case FeatureType.Point:
            filePath = Application.StartupPath + @"\style\td08.style";
```

```csharp
            break;
    }
    string xmlSymbol = File.ReadAllText(filePath);
    IFeatureScheme scheme = xd.Deserialize<IFeatureScheme>(xmlSymbol);
    currentLayer.Symbology = scheme;
}

//保存自定义符号文件
private void btnSaveStyle_Click(object sender, EventArgs e)
{
    SaveFileDialog Savefg = new SaveFileDialog();
    Savefg.AddExtension = true;
    Savefg.Filter = "样式文件|*.style";
    Savefg.DefaultExt = "style";
    if (Savefg.ShowDialog() == System.Windows.Forms.DialogResult.OK)
    {
        IFeatureSymbolizer Symbolizer = _currentLayer.Symbolizer as IFeatureSymbolizer;
        XmlSerializer xs = new XmlSerializer();
        string xmlSymbol = xs.Serialize(Symbolizer);
        string filepath = Savefg.FileName;
        System.IO.File.WriteAllText(filepath, xmlSymbol);
    }
}
```

自定义符号需要本地有符合格式文件的存储路径，在程序的"bin"→"debug"目录下新建"style"文件夹，在此目录下新建三个文本文件，分别命名为"td07""td08""td09"，扩展名自定义为".style"，分别表示线、点、面三种符号类型。分别在三个文件中写入相应的 XML 标记语句，以点符号样式文件"td08"为例说明 XML 文件结构。

点符号样式文件"td08"的根节点和头部代码如下：

```xml
<?xml version="1.0" encoding="utf-16"?>
<root type="0">
  <types>
    <item key="0" value="DotSpatial.Symbology.PointScheme, DotSpatial.Symbology, Version=1.7.0.0, Culture=neutral, PublicKeyToken=6178c08da7998387" />
    <item key="1" value="System.Boolean, mscorlib, Version=4.0.0.0, Culture=neutral, PublicKeyToken=b77a5c561934e089" />
    <item key="2" value="DotSpatial.Symbology.PointCategoryCollection, DotSpatial.Symbology, Version=1.7.0.0, Culture=neutral, PublicKeyToken=6178c08da7998387" />
    <item key="3" value="System.Collections.Generic.List`1[[DotSpatial.Symbology. IPoint-
```

Category, DotSpatial.Symbology, Version=1.7.0.0, Culture=neutral, PublicKeyToken= 6178c08da 7998387]], mscorlib, Version=4.0.0.0, Culture=neutral, PublicKeyToken= b77a5c561934e089" />
        \<item key="4" value="DotSpatial.Symbology.PointCategory, DotSpatial.Symbology, Version=1.7.0.0, Culture=neutral, PublicKeyToken=6178c08da7998387" />
        \<item key="5" value="DotSpatial.Symbology.PointSymbolizer, DotSpatial.Symbology, Version=1.7.0.0, Culture=neutral, PublicKeyToken=6178c08da7998387" />
        \<item key="6" value="DotSpatial.Data.CopyList`1[[DotSpatial.Symbology.ISymbol, DotSpatial.Symbology, Version=1.7.0.0, Culture=neutral, PublicKeyToken=6178c08da7998387]], DotSpatial.Data, Version=1.7.0.0, Culture=neutral, PublicKeyToken=c29dbf30e059ca9d" />
        \<item key="7" value="System.Collections.Generic.List`1 [[DotSpatial. Symbology. ISymbol, DotSpatial.Symbology, Version=1.7.0.0, Culture=neutral, PublicKeyToken= 6178c08da 7998387]], mscorlib, Version=4.0.0.0, Culture=neutral, PublicKeyToken= b77a5c561934e089" />
        \<item key="8" value="DotSpatial.Symbology.SimpleSymbol, DotSpatial.Symbology, Version=1.7.0.0, Culture=neutral, PublicKeyToken=6178c08da7998387" />
        \<item key="9" value="System.String, mscorlib, Version=4.0.0.0, Culture=neutral, PublicKeyToken=b77a5c561934e089" />
        \<item key="10" value="System.Single, mscorlib, Version=4.0.0.0, Culture=neutral, PublicKeyToken=b77a5c561934e089" />
        \<item key="11" value="DotSpatial.Symbology.PointShape, DotSpatial.Symbology, Version=1.7.0.0, Culture=neutral, PublicKeyToken=6178c08da7998387" />
        \<item key="12" value="System.Double, mscorlib, Version=4.0.0.0, Culture=neutral, PublicKeyToken=b77a5c561934e089" />
        \<item key="13" value="DotSpatial.Symbology.Position2D, DotSpatial.Symbology, Version=1.7.0.0, Culture=neutral, PublicKeyToken=6178c08da7998387" />
        \<item key="14" value="DotSpatial.Symbology.Size2D, DotSpatial.Symbology, Version= 1.7.0.0, Culture=neutral, PublicKeyToken=6178c08da7998387" />
        \<item key="15" value="DotSpatial.Symbology.ScaleMode, DotSpatial.Symbology, Version=1.7.0.0, Culture=neutral, PublicKeyToken=6178c08da7998387" />
        \<item key="16" value="System.Drawing.GraphicsUnit, System.Drawing, Version= 4.0.0.0, Culture=neutral, PublicKeyToken=b03f5f7f11d50a3a" />
    \</types>

可见，XML 文件根节点包含版本、字符编码、渲染类型等系统参数。

点符号样式文件"td08"的子节点部分代码如下：
\<member name="IsLegendGroup" type="1" value="False" />
    \<member name="Categories" type="2">
        \<member name="InnerList" type="3">
            \<item type="4">
                \<member name="IsLegendGroup" type="1" value="False" />
                \<member name="Symbolizer" type="5">

```xml
<member name="IsLegendGroup" type="1" value="False" />
<member name="Symbols" type="6">
  <member name="InnerList" type="7">
    <item type="8">
      <member name="XmlColor" type="9" value="#FF0000" />
      <member name="Opacity" type="10" value="1" />
      <member name="PointShapes" type="11" value="Rectangle" />
      <member name="XmlOutlineColor" type="9" value="Black" />
      <member name="OutlineOpacity" type="10" value="1" />
      <member name="OutlineWidth" type="12" value="1" />
      <member name="UseOutline" type="1" value="False" />
      <member name="Angle" type="12" value="0" />
      <member name="Offset" type="13">
        <member name="X" type="12" value="0" />
        <member name="Y" type="12" value="0" />
      </member>
      <member name="Size" type="14">
        <member name="Height" type="12" value="4" />
        <member name="Width" type="12" value="4" />
      </member>
    </item>
  </member>
</member>
<member name="IsVisible" type="1" value="True" />
<member name="ScaleModes" type="15" value="Symbolic" />
<member name="Smoothing" type="1" value="True" />
<member name="Units" type="16" value="Pixel" />
</member>
<member name="SelectionSymbolizer" type="5">
  <member name="IsLegendGroup" type="1" value="False" />
  <member name="Symbols" type="6">
    <member name="InnerList" type="7">
      <item type="8">
        <member name="XmlColor" type="9" value="Cyan" />
        <member name="Opacity" type="10" value="1" />
        <member name="PointShapes" type="11" value="Rectangle" />
        <member name="XmlOutlineColor" type="9" value="Black" />
        <member name="OutlineOpacity" type="10" value="1" />
        <member name="OutlineWidth" type="12" value="1" />
```

```xml
            <member name="UseOutline" type="1" value="False" />
            <member name="Angle" type="12" value="0" />
            <member name="Offset" type="13">
              <member name="X" type="12" value="0" />
              <member name="Y" type="12" value="0" />
            </member>
            <member name="Size" type="14">
              <member name="Height" type="12" value="4" />
              <member name="Width" type="12" value="4" />
            </member>
          </item>
        </member>
      </member>
      <member name="IsVisible" type="1" value="True" />
      <member name="ScaleModes" type="15" value="Symbolic" />
      <member name="Smoothing" type="1" value="True" />
      <member name="Units" type="16" value="Pixel" />
    </member>
    <member name="FilterExpression" type="9" value="[ADCLASS]=1" />
    <member name="LegendText" type="9" value="1" />
  </item>
  </member>
</member>
<member name="AppearsInLegend" type="1" value="True" />
<member name="LegendText" type="9" value="ADCLASS" />
</root>
```

可见，XML 各子节点标记了符号的类型、大小、方向、颜色、轮廓等具体参数。

运行程序，分别加载点、线、面数据，点击"符号渲染"按钮，矢量数据将被渲染成所设置的符号样式。点、线、面数据的运行效果如图 1-3-2～图 1-3-4 所示。

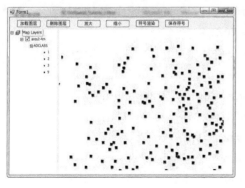

图 1-3-2 点符号唯一值渲染

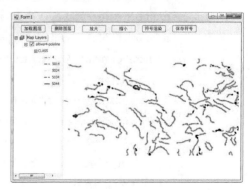

图 1-3-3 线符号唯一值渲染

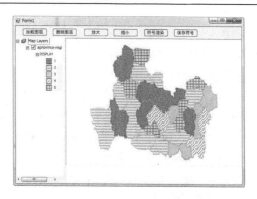

图 1-3-4 面符号唯一值渲染

在左侧的图例控件中双击某个图例对象,将出现符号编辑器窗体。可以对符号进行比例尺模式、符号类型、填充模式、轮廓编辑等操作,以及加入到定制符号。编辑好新符号样式后,回到主窗体点击"保存符号"按钮,可将其保存为新的符号文件。

自主练习:自定义符号,对照编程实现专题地图符号配置。

实验 1-4 地理要素编辑

DotSpatial 支持属性数据的编辑和显示,通过数据表关联可以添加、删除属性字段,通过 IField 接口可对属性字段值进行单个或批量设置和修改,满足属性数据维护的需求。DotSpatial 提供了丰富的空间数据编辑功能,包括点、线、面三种不同数据类型。

(1)实验目的:通过地理要素编辑实习,进一步了解 DotSpatial 主要控件的功能和特性,初步了解空间数据编辑的原理,初步掌握使用 DotSpatial 类库进行图形编辑和属性编辑的方法。

(2)相关实验:实验 1-1 地理数据加载与地图浏览、实验 1-2 地理要素查询与检索。

(3)实验数据:本教材系列实验数据。

(4)实验环境:Visual Studio2010、DotSpatial 1.7 库、.NET Framework 4.0 框架、C#编程语言。

(5)实验内容:通过在 Visual Studio 下使用 Map 控件和代码编写一个窗体应用程序实现对地理要素的属性进行添加、删除、修改等操作的功能;通过在 Visual Studio 下使用 Map 控件和代码编写一个窗体应用程序实现对点、线、面三种类型空间要素进行绘制编辑的功能。

1. 要素属性编辑

拖拽两个 Button 按钮到主窗体中,分别命名为"btnAddField""btnDeleteField",按钮文本为"添加字段""删除字段",在各个按钮的事件中编写相应的代码。

实验代码如下:

```
using DotSpatial.Controls;
using DotSpatial.Data;
//添加属性字段
private void btnAddField_Click(object sender, EventArgs e)
```

```csharp
        {
            System.Data.DataTable dt = null;
            if (map1.Layers.Count > 0)
            {
                IMapFeatureLayer mapLayer = default(IMapFeatureLayer);
                mapLayer = map1.Layers[0] as IMapFeatureLayer;
                dt = mapLayer.DataSet.DataTable;
                DataColumn column = new DataColumn("People");
                dt.Columns.Add(column);
                mapLayer.DataSet.Save();
            }
            else
            {
                MessageBox.Show("图层不存在.");
            }
        }
        //删除属性字段
        private void btnDeleteField_Click(object sender, EventArgs e)
        {
            System.Data.DataTable dt = null;
            if (map1.Layers.Count > 0)
            {
                IMapFeatureLayer mapLayer = default(IMapFeatureLayer);
                mapLayer = map1.Layers[0] as IMapFeatureLayer;
                dt = mapLayer.DataSet.DataTable;
                dt.Columns.Remove("People");
                mapLayer.DataSet.Save();
            }
            else
            {
                MessageBox.Show("图层不存在.");
            }
        }
```

运行程序，加载地图，点击"添加字段"按钮，再点击"选择图斑"，鼠标点击地图某个图斑，弹出属性表。添加字段效果如图 1-4-1 所示。属性表最右侧多出一列名为"People"的属性列，可进行属性值编辑。点击"删除字段"按钮可删除这列属性，删除字段效果如图 1-4-2 所示。

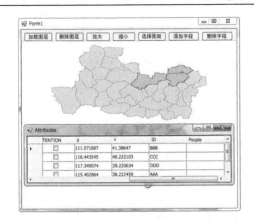
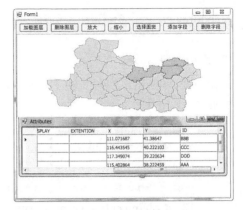

图 1-4-1　添加属性字段界面　　　　图 1-4-2　删除属性字段界面

2. 创建空间图形要素

新建空间图形需要设置要素属性，而编辑（增加、修改、删除）空间图形则需要设置要素捕捉，结合鼠标操作可实现矢量图形自定义绘制和编辑。

1）新建与保存要素

拖拽六个 Button 按钮到主窗体中，分别命名为"btnCreatePoint""btnCreateLine""btnCreatePolygon""btnSavePoint""btnSaveLine""btnSavePolygon"，按钮文本为"新建点要素""新建线要素""新建面要素""保存点要素""保存线要素""保存面要素"，在各个按钮的事件中编写相应的代码。

各事件的实验代码如下：

```
using DotSpatial.Controls;
using DotSpatial.Symbology;
using DotSpatial.Data;
using DotSpatial.Topology;
　//点要素属性变量
　string shapeType;
　FeatureSet pointF = new FeatureSet(FeatureType.Point);
　int pointID = 0;
　bool pointmouseClick = false;
//线要素属性变量
　MapLineLayer lineLayer = default(MapLineLayer);
　FeatureSet lineF = new FeatureSet(FeatureType.Line);
　int lineID = 0;
　bool firstClick = false;
　bool linemouseClick = false;
//面要素属性变量
　FeatureSet polygonF = new FeatureSet(FeatureType.Polygon);
　int polygonID = 0;
```

```csharp
        bool polygonmouseClick = false;

//新建点要素方法
    private void btnCreatePoint_Click(object sender, EventArgs e)
    {
        map1.Cursor = Cursors.Cross;
        shapeType = "Point";
        //set projection
        pointF.Projection = map1.Projection;
        DataColumn column = new DataColumn("PointID");
        pointF.DataTable.Columns.Add(column);
        MapPointLayer pointLayer = (MapPointLayer)map1.Layers.Add(pointF);
        PointSymbolizer symbol = new PointSymbolizer(Color.Red,
        DotSpatial.Symbology.PointShape.Ellipse, 3);
        pointLayer.Symbolizer = symbol;
        pointLayer.LegendText = "point";
        pointmouseClick = true;
    }
//新建线要素方法
    private void btnCreateLine_Click(object sender, EventArgs e)
    {
        map1.Cursor = Cursors.Cross;
        shapeType = "line";
        lineF.Projection = map1.Projection;
        DataColumn column = new DataColumn("LineID");
        if (!lineF.DataTable.Columns.Contains("LineID"))
        {
            lineF.DataTable.Columns.Add(column);
        }
        lineLayer = (MapLineLayer)map1.Layers.Add(lineF);
        LineSymbolizer symbol = new LineSymbolizer(Color.Blue, 2);
        lineLayer.Symbolizer = symbol;
        lineLayer.LegendText = "line";
        firstClick = true;
        linemouseClick = true;
    }
//新建面要素方法
    private void btnCreatePolygon_Click(object sender, EventArgs e)
    {
```

```csharp
    map1.Cursor = Cursors.Cross;
    shapeType = "polygon";
    polygonF.Projection = map1.Projection;
    DataColumn column = new DataColumn("PolygonID");
    if (!polygonF.DataTable.Columns.Contains("PolygonID"))
    {
        polygonF.DataTable.Columns.Add(column);
    }
    MapPolygonLayer polygonLayer = (MapPolygonLayer)map1.Layers.Add(polygonF);
    PolygonSymbolizer symbol = new PolygonSymbolizer(Color.Green);
    polygonLayer.Symbolizer = symbol;
    polygonLayer.LegendText = "polygon";
    firstClick = true;
    polygonmouseClick = true;
}

//保存点要素方法
private void btnSavePoint_Click(object sender, EventArgs e)
{
    pointF.SaveAs(Application.StartupPath + @"\shp\"+"point.shp", true);
    MessageBox.Show("保存点要素成功！");
    map1.Cursor = Cursors.Arrow;
}
//保存线要素方法
private void btnSaveLine_Click(object sender, EventArgs e)
{
    lineF.SaveAs(Application.StartupPath + @"\shp\" + "line.shp", true);
    MessageBox.Show("保存线要素成功！");
    map1.Cursor = Cursors.Arrow;
    linemouseClick = false;
}
//保存面要素方法
private void btnSavePolygon_Click(object sender, EventArgs e)
{
    polygonF.SaveAs(Application.StartupPath + @"\shp\" + "polygon.shp", true);
    MessageBox.Show("保存面要素成功！");
    map1.Cursor = Cursors.Arrow;
    polygonmouseClick = false;
}
```

2）绘制要素

右击 Map 控件，在属性窗体中双击"MouseDown"事件，编写通过鼠标操作实现绘制点、线、面三种要素的事件。

绘制要素事件的实验代码如下：

```csharp
//地图控件绑定鼠标按下事件
private void map1_MouseDown(object sender, MouseEventArgs e)
{
    switch (shapeType)
    {
        case "Point":
            if (e.Button == MouseButtons.Left)
            {
                if ((pointmouseClick))
                {
                    Coordinate coord = map1.PixelToProj(e.Location);
                    DotSpatial.Topology.Point point = new
                    DotSpatial.Topology.Point(coord);
                    IFeature currentFeature = pointF.AddFeature(point);
                    pointID = pointID + 1;
                    map1.ResetBuffer();
                }
                else
                {
                    map1.Cursor = Cursors.Default;
                    pointmouseClick = false;
                }
            }
            break;
        case "line":
            if (e.Button == MouseButtons.Left)
            {
                Coordinate coord = map1.PixelToProj(e.Location);
                if (linemouseClick)
                {
                    if (firstClick)
                    {
                        List<Coordinate>lineArray = new List<Coordinate>();
                        LineString lineGeometry = new LineString(lineArray);
                        IFeature lineFeature =lineF.AddFeature(lineGeometry);
```

```csharp
                        lineFeature.Coordinates.Add(coord);
                        lineID = lineID + 1;
                        lineFeature.DataRow["LineID"] = lineID;
                        firstClick = false;
                    }
                    else
                    {
                        IFeature existingFeature=lineF.Features[lineF.Features.Count-1];
                        existingFeature.Coordinates.Add(coord);
                        if (existingFeature.Coordinates.Count >= 2)
                        {
                            lineF.InitializeVertices();
                            map1.ResetBuffer();
                        }
                    }
                }
            }
            else
            {
                firstClick = true;
                map1.ResetBuffer();
            }
            break;
        case "polygon":
            if (e.Button == MouseButtons.Left)
            {
                Coordinate coord = map1.PixelToProj(e.Location);
                if (polygonmouseClick)
                {
                    if (firstClick)
                    {
                        List<Coordinate>polygonArray=new List <Coordinate>();
                        LinearRing polygonGeometry = new LinearRing
                        (polygonArray);
                        IFeature polygonFeature = polygonF.
                        AddFeature(polygonGeometry);
                        polygonFeature.Coordinates.Add(coord);
                        polygonID = polygonID + 1;
                        polygonFeature.DataRow["PolygonID"] = polygonID;
```

```
                        firstClick = false;
                    }
                    else
                    {
                        IFeature existingFeature =(IFeature)polygonF.
                        Features[polygonF.Features.Count - 1];
                        existingFeature.Coordinates.Add(coord);
                        if (existingFeature.Coordinates.Count >= 3)
                        {
                            polygonF.InitializeVertices();
                            map1.ResetBuffer();
                        }
                    }
                }
            }
            else
            {
                firstClick = true;
            }
            break;
        }
    }
```

运行程序，加载点数据，点击"新建点要素"按钮，点击地图鼠标出现十字样式并绘制点，再次点击地图可绘制新点。类似地，线要素和面要素也可通过移动鼠标绘制，并进行保存操作。在地图上绘制点、线、面三种要素的效果如图 1-4-3～图 1-4-5 所示。

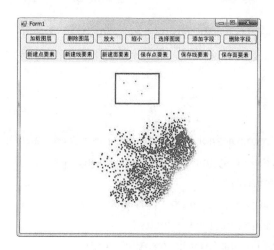
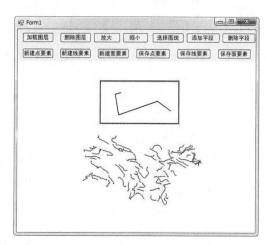

图 1-4-3　绘制点要素界面　　　　　　　图 1-4-4　绘制线要素界面

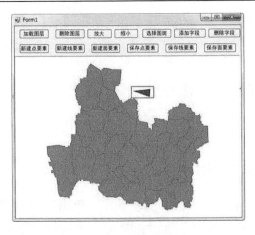 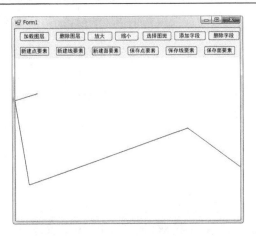

图 1-4-5　绘制面要素界面　　　　图 1-4-6　显示保存的线要素界面

（包括当前绘制与本地保存）

点击"保存点要素""保存线要素""保存面要素"按钮将保存当前绘制的各类型数据，其中加载本地保存的线要素后的效果如图 1-4-6 所示。

自主练习：对照编程，实现图形新建、属性编辑的功能。

实验 1-5　空间分析与统计

DotSpatial 的 Topology 类库和 Tool 类库中集成了很多空间叠置分析方法，其中叠加裁切是空间数据分析中常用的方法。通过底图和叠加图进行叠置分析，可生成叠加后的新图层，其属性集成了底图和叠加图的属性。叠加图可以是点、线、面图层，而底图必须是面图层。

DotSpatial 的 Data 类库集成了一些空间统计方法，IFeature 接口下的 Area 方法实现了统计每个矢量图斑面积的功能。

（1）实验目的：通过空间分析与统计实习，进一步了解 DotSpatial 主要控件的功能和特性，初步了解空间分析与统计的原理，初步掌握使用 DotSpatial 类库进行空间叠加和面积统计分析等方法。

（2）相关实验：实验 1-1 地理数据加载与地图浏览、实验 1-2 地理要素查询与检索。

（3）实验数据：本教材系列实验数据。

（4）实验环境：Visual Studio2010、DotSpatial 1.7 库、.NET Framework 4.0 框架、C# 编程语言。

（5）实验内容：通过在 Visual Studio 下使用 Map 控件和代码编写一个窗体应用程序实现对矢量数据进行叠加裁切分析的功能；通过在 Visual Studio 下使用 Map 控件和代码编写一个窗体应用程序实现统计矢量数据图斑面积的功能。

1. 叠加裁切分析

在项目名上右击，添加一个新的 WinFrom 窗体，命名为"Clip"，添加相应事件，如图 1-5-1 所示。

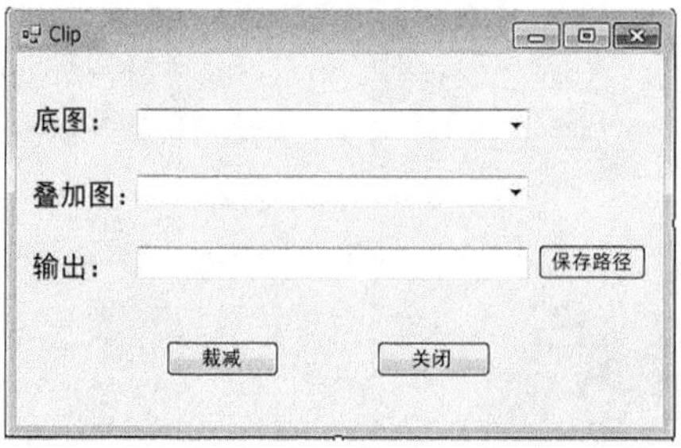

图 1-5-1　叠加裁切设置界面

从工具箱中拖拽三个 Label 文本标签，内容分别为"底图""叠加图""输出"；拖拽两个 ComboBox 下拉框和一个 TextBox 文本框，分别命名为"Basemaplayer""overlaymaplayer""OutLayerPath"；再拖拽三个 Button 按钮，分别命名为"btnSelectLayerPath""btnClip""btnClose"；文本内容为"保存路径""裁减""关闭"；然后拖拽一个 backgroundWorker1 到窗体上用于后台多线程执行分析操作，在此窗体下编写相应代码。

窗体事件的实验代码如下：

```
using DotSpatial.Controls;
using DotSpatial.Data;
using DotSpatial.Tools;
using DotSpatial.Topology

    //窗体初始化绑定图层事件
    private void Clip_Load(object sender, EventArgs e)
    {
        Basemaplayer.Items.Clear();
        overlaymaplayer.Items.Clear();
        for (int i = 0; i < _map.Layers.Count; i++)
        {
            Basemaplayer.Items.Add(_map.Layers[i].DataSet.Name);
            overlaymaplayer.Items.Add(_map.Layers[i].DataSet.Name);
        }
        Basemaplayer.SelectedIndex = 0;
        overlaymaplayer.SelectedIndex = 1;
    }
    //后台工作时调用叠加处理方法
    private void backgroundWorker1_DoWork(object sender, DoWorkEventArgs e)
    {
        DotSpatial.Modeling.Forms.ToolProgress frmProgress = (e.Argument
        as object[])[0] as DotSpatial.Modeling.Forms.ToolProgress;
        IFeatureSet baseData = (e.Argument as object[])[1] as IFeatureSet;
```

```csharp
    if (baseData.FeatureType != DotSpatial.Topology.FeatureType.Polygon)
    {
        MessageBox.Show("底图必须是面图层,请重新选择底图!");
        return;
    }

    IFeatureSet clipData = (e.Argument as object[])[2] as IFeatureSet;
    if (clipData.FeatureType != DotSpatial.Topology.FeatureType.Line &&
        clipData.FeatureType != DotSpatial.Topology.FeatureType.Polygon)
    {
        MessageBox.Show("裁剪图必须是面图层或线图层,请重新选择底图!");
        return;
    }

    IFeatureSet outPutData = new FeatureSet(DotSpatial.Topology.FeatureType.Polygon);
    outPutData.Filename = (e.Argument as object[])[3] as string;

    switch (clipData.FeatureType)
    {
        case DotSpatial.Topology.FeatureType.Polygon:
            ClipPolygonWithPolygon clipByPolygon = new ClipPolygonWithPolygon();
            clipByPolygon.Execute(baseData, clipData, outPutData, frmProgress);
            break;
        case DotSpatial.Topology.FeatureType.Line:
            ClipPolygonWithLine clipByLine = new ClipPolygonWithLine();
            clipByLine.Execute(baseData, clipData, outPutData, frmProgress);
            break;
    }
}
//后台工作完成消息处理
private void backgroundWorker1_RunWorkerCompleted(object sender,
RunWorkerCompletedEventArgs e)
{
    if (MessageBox.Show("裁剪已完成,是否加载裁剪后的文件?", "提示:",
    MessageBoxButtons.YesNo) != System.Windows.Forms.DialogResult.Yes) return;
    IFeatureSet outPutData = FeatureSet.Open(this.OutLayerPath.Text);
    _map.Layers.Clear();
```

```csharp
            _map.Layers.Add(outPutData);
        }
        //执行叠加裁切方法
        private void btnClip_Click(object sender, EventArgs e)
        {
            DotSpatial.Modeling.Forms.ToolProgress frmProgress = new DotSpatial.
            Modeling.Forms.ToolProgress(1);
            this.backgroundWorker1.RunWorkerAsync(new object[4] {
                frmProgress,
                _map.Layers[this.Basemaplayer.SelectedIndex].DataSet,
                _map.Layers[this.overlaymaplayer.SelectedIndex].DataSet,
                this.OutLayerPath.Text
            });
            this.Close();
        }
```

在主窗体上拖拽一个 Button 按钮，命名为"btnClip"，文本内容为"裁切"，在按钮下编写相应代码用于打开"Clip"窗体。

```csharp
        //打开裁切窗体事件
        private void btnClip_Click(object sender, EventArgs e)
        {
            Clip frmClip = new Clip(this.map1);
            frmClip.ShowDialog();
        }
```

运行程序，先后添加两个图层，点击"裁切"按钮，弹出裁切窗体，此时底图和叠加图分别出现加载好的图层名称，点击"保存路径"，在保存对话框中选择要输出图层的路径。参数设置好后点击"裁切"按钮，经过叠加处理后，将生成裁切后的图层。

自主练习：对照编程实现图形裁切功能。

2. 面积统计分析

拖拽一个 Button 按钮到主窗体中，命名为"btnTotalArea"，按钮文本为"all"；添加一个 ComboBox 下拉列表到主窗体中，命名为"comboBoxName"；再从工具箱中拖拽三个 Label 文本标签，内容置空；在按钮和下拉列表的事件中编写相应的代码。

实验代码如下：

```csharp
        //统计全部图斑面积
        private double getTotalArea(DotSpatial.Controls.Map mapInput)
        {
            double stateArea = 0;
            if ((mapInput.Layers.Count > 0))
            {
                MapPolygonLayer stateLayer = default(MapPolygonLayer);
```

```csharp
            stateLayer = (MapPolygonLayer)mapInput.Layers[0];
            if ((stateLayer == null))
            {
                MessageBox.Show("图层必须是面图层！");
            }
            else
            {
                foreach (IFeature stateFeature in stateLayer.DataSet.Features)
                {
                    stateArea += stateFeature.Area();
                }
            }
        }
        return stateArea;
    }

//统计选择属性字段的图斑面积
private double getSelectArea(string uniqueColumnName, string uniqueValue, DotSpatial.Controls.Map mapInput)
{
    double stateArea = 0;
    if ((mapInput.Layers.Count > 0))
    {
            MapPolygonLayer stateLayer = default(MapPolygonLayer);
            stateLayer = (MapPolygonLayer)mapInput.Layers[0];
            if ((stateLayer == null))
            {
                MessageBox.Show("图层必须是面图层！");
            }
            else
            {
                stateLayer.SelectByAttribute("[" + uniqueColumnName + "] =" + "'" + uniqueValue + "'");
                foreach (IFeature stateFeature in stateLayer.DataSet.Features)
                {
                    if (uniqueValue.CompareTo(stateFeature.DataRow[uniqueColumnName]) == 0)
                    {
                        stateArea = stateFeature.Area();
```

```
                    return stateArea;
                }
            }
        }
    }
    return stateArea;
}

//统计全部图斑面积按钮事件
private void btnTotalArea_Click(object sender, EventArgs e)
{
    laberTotalArea.Text = getTotalArea(map1).ToString()+"万平方公里";
}

//选择下拉列表的属性字段来统计相应图斑面积事件
private void comboBoxName_SelectedIndexChanged(object sender, EventArgs e)
{
    laberSelectArea.Text = getSelectArea("NAME", comboBoxName.SelectedItem.
    ToString(), map1).ToString() + "万平方公里";
}
```

运行程序，加载地图，点击"all"按钮，按钮右边将显示出整体面积值；从下拉列表中可选择任一图斑，显示其面积。

自主练习：对照编程实现由图形面积统计的地理信息系统功能。

实验 1-6　地图打印输出

DotSpatial 可以将当前处理的矢量图层输出为工程文件，文件格式为 dspx，这样可以及时保存地图操作记录，方便用户随时处理地图。

DotSpatial 通过 LayoutControl 控件集成了较为丰富的地图整饰功能，用户可以通过鼠标操作直接生成比例尺、图例、指北针、文本等地图输出要素，还可以对地图及要素进行放大、缩小、平移等基本操作。结合本地打印机可输出自定义地图。

（1）实验目的：通过地图打印输出实习，进一步了解 DotSpatial 主要控件的功能和特性，初步了解地图输出和打印的原理，初步掌握使用 DotSpatial 类库进行地图工程文件的输出和地图布局整饰打印的方法。

（2）相关实验：实验 1-1 地理数据加载与地图浏览、实验 1-2 地理要素查询与检索。

（3）实验数据：本教材系列实验数据。

（4）实验环境：Visual Studio2010、DotSpatial 1.7 库、.NET Framework 4.0 框架、C# 编程语言。

（5）实验内容：通过在 Visual Studio 下使用 Map 控件、AppManager 组件和代码编写一个窗体应用程序,实现输出地图工程文件的功能;通过在 Visual Studio 下使用 Map 控件、LayoutControl 控件和代码编写一个窗体应用程序,实现整饰地图布局并进行打印输出的功能。

1. 地图工程文件输出

从工具箱中拖拽"AppManager"组件到窗体中,再拖拽两个 Button 按钮到主窗体中,分别命名为"btnAddProject""btnSaveProject",按钮文本为"导入工程""保存工程",在各个按钮的事件中编写相应的代码。

实验代码如下:

```csharp
using DotSpatial.Controls;
//导入工程文件
private void btnAddProject_Click(object sender, EventArgs e)
{
    OpenFileDialog openfg = new OpenFileDialog();
    openfg.AddExtension = true;
    openfg.Filter = "工程文件|*.dspx";
    openfg.DefaultExt = "dspx";
    if (openfg.ShowDialog() == System.Windows.Forms.DialogResult.OK)
    {
        appManager1.Map = map1;
        appManager1.SerializationManager.OpenProject(openfg.FileName);
    }
}

//保存工程文件
private void btnSaveProject_Click(object sender, EventArgs e)
{
    SaveFileDialog Savefg = new SaveFileDialog();
    Savefg.AddExtension = true;
    Savefg.Filter = "工程文件|*.dspx";
    Savefg.DefaultExt = "dspx";
    if (Savefg.ShowDialog() == System.Windows.Forms.DialogResult.OK)
    {
        appManager1.Map = map1;
        appManager1.SerializationManager.SaveProject(Savefg.FileName);
    }
}
```

运行程序,加载地图,导入地图后对图层进行随机颜色渲染,点击"保存工程"按钮,在对话框中选择保存路径,工程将保存到本地;点击"导入工程"按钮,打开保存的工程文件,将直接出现渲染后的地图,如图 1-6-1 所示。

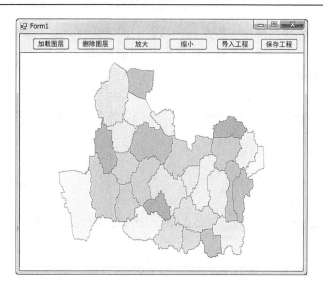

图 1-6-1　地图工程文件渲染输出图

2. 地图整饰打印

添加一个新的 WinFrom 窗体，命名为"LayoutForm"，从工具箱中拖拽"LayoutControl"、"LayoutListBox"和"LayoutPropertyGrid"三个控件到窗体中，将"LayoutControl"控件的"LayoutListBox"和"LayoutPropertyGrid"属性关联到另外两个控件；拖拽一个"ToolStrip"菜单栏，右击添加十个按钮；再拖拽一个 Panel（面板）到窗体中，在其中加入七个 Button（按钮）；最后在各个按钮的事件中编写相应的代码。

打印输出窗体设计界面如图 1-6-2 所示。

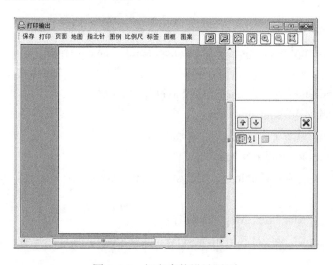

图 1-6-2　打印窗体设计界面

窗体事件的代码如下：

```
using DotSpatial.Controls;
//打印窗体初始化设置
```

```csharp
private void LayoutForm_Load(object sender, EventArgs e)
{
    MapControl.ZoomToMaxExtent();
    layoutControl1.ZoomFitToScreen();
    if (MapControl != null)
    {
        var mapElement = layoutControl1.CreateMapElement();
        mapElement.Size = layoutControl1.Size;
        layoutControl1.AddToLayout(mapElement);
    }
}
//保存打印视图
private void btnSaveAs_Click(object sender, EventArgs e)
{
    layoutControl1.SaveLayout(true);
}
//打印视图
private void btnPrint_Click(object sender, EventArgs e)
{
    layoutControl1.Print();
}
//打印页面设置
private void btnPageSetup_Click(object sender, EventArgs e)
{
    layoutControl1.ShowPageSetupDialog();
}
//添加打印地图
private void btnMap_Click(object sender, EventArgs e)
{
    if (_layoutControl1.MapControl != null)
    {
        layoutControl1.AddElementWithMouse(_layoutControl1.CreateMapElement());
    }
    else
    {
        MessageBox.Show(Parent, "添加地图失败.", "错误",MessageBoxButtons.OK,
        MessageBoxIcon.Information);
    }
}
```

```csharp
//添加指北针
  private void btnNorth_Click(object sender, EventArgs e)
  {
      layoutControl1.AddElementWithMouse(new LayoutNorthArrow());
  }
//添加图例
  private void btnLegend_Click(object sender, EventArgs e)
  {
      layoutControl1.AddElementWithMouse(_layoutControl1.CreateLegendElement());
  }
//添加比例尺
  private void btnScalebar_Click(object sender, EventArgs e)
  {
      layoutControl1.AddElementWithMouse(_layoutControl1.CreateScaleBarElement());
  }
//添加文本标签
  private void btnText_Click(object sender, EventArgs e)
  {
      layoutControl1.AddElementWithMouse(new LayoutText());
  }
//添加图框
  private void btnRectangle_Click(object sender, EventArgs e)
  {
      layoutControl1.AddElementWithMouse(new LayoutRectangle());
  }
//添加图片
  private void btnBitmap_Click(object sender, EventArgs e)
  {
      var ofd = new OpenFileDialog
      {
          Filter = "Images (*.png,*.jpg, *.bmp,*.gif,*.tif)|*.png;*.jpg;*. bmp;*.gif;*. tif",
          FilterIndex = 1,
          CheckFileExists = true
      };
      if (ofd.ShowDialog(Parent) == DialogResult.OK)
      {
          var newBitmap = new LayoutBitmap { Size = new SizeF(100, 100), Filename= ofd.FileName };
          layoutControl1.AddElementWithMouse(newBitmap);
```

```csharp
        }
    }
//放大地图
    private void btnZoomIn_Click(object sender, EventArgs e)
    {
        layoutControl1.ZoomInMap();
    }
//缩小地图
    private void ZoomOut_Click(object sender, EventArgs e)
    {
        layoutControl1.ZoomOutMap();
    }
//居中显示地图
    private void ZoomExtent_Click(object sender, EventArgs e)
    {
        try
        {
            layoutControl1.ZoomFullViewExtentMap();
        }
        catch
        {
            MessageBox.Show("全屏显示地图有误！", "错误", MessageBoxButtons.OK,
            MessageBoxIcon.Warning);
        }
    }
//平移地图
    private void btnPan_Click(object sender, EventArgs e)
    {
        try
        {
            layoutControl1.MapPanMode = true;
        }
        catch
        {
            MessageBox.Show("你移动的地图有误！", "错误", MessageBoxButtons.OK,
            MessageBoxIcon.Warning);
        }
    }
//放大视图窗口
```

```csharp
private void _btnZoomIn_Click(object sender, EventArgs e)
{
    layoutControl1.ZoomIn();
}
//缩小视图窗口
private void _btnZoomOut_Click(object sender, EventArgs e)
{
    layoutControl1.ZoomOut();
}
//居中显示视图窗口
private void _btnZoomExtent_Click(object sender, EventArgs e)
{
    layoutControl1.ZoomFitToScreen();
}
```

在主窗体上添加一个 Button 按钮,命名为"btnPrint",文本内容为"打印输出",执行弹出打印布局窗体,在按钮的事件中编写相应代码。

打印输出的参照代码如下:

```csharp
//弹出打印布局窗体事件
private void printBtn_Click(object sender, EventArgs e)
{
    try
    {
        LayoutForm layout = new LayoutForm();
        layout.MapControl = map2 as Map;
        layout.Show();
    }
    catch
    {
        MessageBox.Show("没有安装打印机!", "错误", MessageBoxButtons.OK,
        MessageBoxIcon.Warning);
    }
}
```

运行程序,加载一个地图并进行符号渲染。点击"打印输出"按钮,弹出打印布局窗体,地图将被加载到打印窗体中,如图 1-6-3 所示。

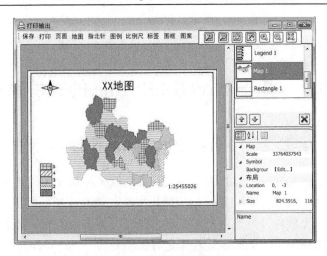

图 1-6-3 打印地图要素布局界面

点击"页面",可对视图进行页面设置,设置张纸大小为"A4",纸张方向为"横向",点击确定视图大小方向发生改变;分别点击"指北针""图例""比例尺""标签"等按钮添加地图要素,通过右侧导航栏可改变不同要素所在图层顺序,根据要素属性栏可对不同要素的大小、颜色、位置等参数进行设置;点击右上方工具栏中的各个按钮,可缩放、平移、居中显示地图及要素。设置完地图布局后点击"打印"可输出当前视图,点击"保存"可随时保存当前视图。

自主练习:对照编程,实现地图配置、打印输出功能。

第二部分 基于 ArcGIS Engine 的开发系列实验

实验 2-1 ArcGIS Engine 控件的使用

ArcGIS Engine 是 ArcObjects 的子集，能够完整地嵌入 GIS 组件库和工具，主要用于开发独立的 GIS 应用程序，或者在自定义的应用程序或已经存在的应用程序中嵌入 GIS 功能。它由两部分组成：一个软件开发包（SDK）和一个可分发的运行时（Runtime）。除此之外，应用程序的功能可以使用可选的扩展模块来进行扩展。

（1）实验目的：通过学习使用 ArcGIS Engine 的基本控件，掌握快速搭建可以独立运行的 GIS 应用程序的能力。

（2）相关实验：GIS 专业实验设备与环境配置中的"GIS 应用开发环境"和"GIS 应用开发资源"。

（3）实验数据：ArcGIS Engine 自带的示例数据或本教材系列实验数据。

（4）实验环境：Visual Studio 2012、ArcGIS Engine 10.2 和 C#语言。

（5）实验内容：各控件的功能属性认知；使用控件搭建简单的 GIS 使用程序。

1. 控件初识

ArcGIS Engine 为开发者提供了可视化的控件，包括：LicenseControl、MapControl、PageLayoutControl、TocControl、ToolbarControl、SceneControl、GlobeControl、SymbologyControl、ArcReaderControl 和 ArcReaderGlobeControl。

在 VS.NET 中通过 ESRI interop 程序集使用 ArcObjects，ESRI interop 程序集为 ArcGIS 控件提供了能够位于.NET 窗体上的控件（继承自 AxHost 类），其控件名前缀为"Ax"，如 AxMapControl、AxPageLayoutControl、AxTOCControl 和 AxToolbarControl 等。每个 ArcGIS Engine 控件都具有方法、属性与事件，它们能够被控件嵌入的容器（如.NET 窗体）访问。每个控件对象及其功能可以与其他 ESRI ArcObjects 和自定义控件组合使用，创建用户化的客户应用程序。需要指出的是，基于 ArcGIS Engine 10.2 构建应用程序时，最好选择.NET Framework 4.5。使用 ArcGIS Engine 控件时，引用库需含有 ESRI.ArcGIS.Controls 和 ESRI.ArcGIS.AxControls，这两个库的命名空间均为 ESRI.ArcGIS.Controls，为此所有使用控件的类中需要在开始处写 using ESRI.ArcGIS.Controls。

2. LicenseControl 控件

LicenseControl 控件用于初始化应用程序时，需提供合适的功能授权；该控件在可视化环境设计时可见，在运行时不可见。在设计阶段，可以使用 LicenseControl 属性页为应用程序配置产品和扩展许可；在运行阶段，如果需要对许可进行调整，应当使用 ArcGIS Engine 的 System 库中的 AoInitalize 对象通过代码进行产品绑定和许可修改。

在调用任何 ArcObjects 代码（包括许可证初始化）之前，应用程序必须找到已安装的相应 ArcGIS 产品。当 API 通过 ESRI.ArcGIS.Version 命名空间中的 ESRI.ArcGIS.RuntimeManager

类来绑定相应的 ArcGIS 运行时，LicenseControl 控件属性选项卡中的 Products 部分，需至少勾选其中之一，默认情况下选择的是 ArcGIS Engine 许可；Extensions 部分，用于选择与 Products 中已选择许可的相应扩展许可，当使用 SceneControl 或 GlobeControl 时，必须选择 3D Analyst 扩展许可；Shutdown 部分，在勾选该选择框的情况下，LicenseControl 初始化许可失败时会自动关闭应用程序。除此之外，还需要在 Program.cs 类的应用程序主入口处添加以下代码来明确绑定的产品：

ESRI.ArcGIS.RuntimeManager.Bind(ESRI.ArcGIS.ProductCode.Engine);

以上代码应当在 Application.Run(new Main_Form()); 这句代码之前添加，否则视为绑定产品失败。另外还可以使用"项目"→"Add ArcGIS License Checking…"选择产品许可并绑定产品。此时会在项目中自动生成 LicenseIntializer.cs 类来辅助绑定产品许可。

在 LicenseIntializer.cs 类中，引用库需含有 ESRI.ArcGIS.Version 库，该 Version 库的命名空间为 ESRI.ArcGIS，为此在 LicenseIntializer.cs 类的开始部分需要写 using ESRI.ArcGIS；该命名空间中 ProductCode 为枚举类型，除 Engine 外还有 ArcReader、Desktop、EngineorDesktop 和 Server 共五个类型；RuntimeManager 为公有静态类，使用 Bind 函数或 BindLicense 函数来绑定许可。

在 Program.cs 类中，使用 LicenseIntializer.cs 类实例化的对象绑定产品和许可。引用库需含有 ESRI.ArcGIS.System，该 System 库的命名空间为 ESRI.ArcGIS.esriSystem，为此在 LicenseIntializer.cs 类的开始部分需要写 using ESRI.ArcGIS.esriSystem；该命名空间中 esriLicenseProductCode 和 esriLicenseExtensionCode 是枚举类型，用于提供产品和许可类型。

自主练习：分别通过手动和"Add ArcGIS License Checking…"的方式绑定产品和许可，根据两者的区别与联系，理解绑定产品和许可的过程及它们的依赖关系；对照 OMD（object model diagrams）尝试调用 LicenseControl 的属性和方法。

3. MapControl 控件

MapControl 控件对应 ArcMap 的数据视图，并封装了 Map 对象。MapControl 可以加载地图文档、添加删除图层、与键盘和鼠标交互操作等。详细可参考 ControlsObjectModel.pdf 和帮助文档。

1）MapControl 属性

使用 ArcMap 预先构建的地图文档可以加载到 MapControl 中，从而无须以编程方式进行地图可视化。在 MapControl 属性页可以定制外观并可以设置加载地图文档，有链接型和完全读取型两种形式。设置链接型时，MapControl 读取地图文档，并显示地图文档的最新更新。设置完全读取型时，MapControl 会将地图文档的内容复制到 MapControl 中，但不会显示从该点开始对地图文档所做的进一步更新。

2）加载地图文档

MapControl 也通过 CheckMxFile 方法确定文档是否有效，使用 LoadMxFile 方法将地图文档加载到 MapControl 中。使用 LoadMxFile 的方法需要交互操作，可以采用 VS.NET 自带的 MenuStrip 进行，添加打开地图文档的方法作为事件触发。另外需要 VS.NET 平台 System.Windows.Forms 库中的 OpenFileDialog 进行文件选择，只需将 OpenFileDialog 得到的文件路径赋值给 CheckMxFile 和 LoadMxFile 即可，详细代码可联系作者索取。

3）MapControl 控件交互操作

地图控件与鼠标和键盘交互是常用的操作，包括移动地图、改变地图显示范围和绘制图形等。

（1）箭头键移动地图。使用箭头键移动地图需要设置 MapControl 的两个属性：AutoKeyboardScrolling 和 KeyIntercept。其中 AutoKeyboardScrolling 为 Bool 型，设置为 true；KeyIntercept 为 Int 型（OMD 文件中此处错标为 Long 型），其值对应 SystemUI 库中枚举类型 esriKeyIntercept 的值，应当设置为 1 或者 esriKeyInterceptArrowKeys，见表 2-1-1。往往在程序加载时就设置箭头键是否可以移动地图，为此本书将该部分代码添加在 Form_Load 事件中。因为使用了 SystemUI 库的枚举类型 esriKeyIntercept，所以在设置箭头键移动地图的类中应当写 using ESRI.ArcGIS.SystemUI。

表 2-1-1　枚举类型 esriKeyIntercept 的值

枚举常量	值	描述
esriKeyInterceptNone	0	No keys are intercepted
esriKeyInterceptArrowKeys	1	Intercepts the arrow keys, normally handled by the container to change control focus
esriKeyInterceptAlt	2	Intercepts the Alt key, normally handled by a container to change focus
esriKeyInterceptTab	4	Intercepts the Tab key, normally handled by the container to change control focus
esriKeyInterceptEnter	8	Intercepts the Enter key, normally handled by the container to click the default button

（2）鼠标平移旋转地图。鼠标平移地图时会触发鼠标按键按下、鼠标平移和鼠标按键抬起的事件，而地图平移是一系列移动地图的操作。MapControl 封装了 Pan 函数用于地图平移，只需要某一事件触发 Pan 函数，后续的平移地图操作就都由 Pan 函数完成。在平移地图时往往会改变鼠标的形状，可以使用 MapControl 的 MousePointer 来修改，其对应的是 Controls 库的枚举类型 esriControlsMousePointer 的值，其值为：-1-esriPointerParentWindow、0-esriPointerDefault、1-esriPointerArrow、2-esriPointerCrosshair、3-esriPointerIBeam、4-esriPointerIcon、5-esriPointerSize、6-esriPointerSizeNESW、7-esriPointerSizeNS、8-esriPointerSizeNWSE、9-esriPointerSizeWE、10-esriPointerUpArrow、11-esriPointerHourglass、12-esriPointerNoDrop、13-esriPointerArrowHourglass、14-esriPointerArrowQuestion、15-esriPointerSizeAll、50-esriPointerZoom、51-esriPointerZoomIn、52-esriPointerZoomOut、53-esriPointerPan、54-esriPointer Panning、55-esriPointerIdentify、56-esriPointerLabel、57-esriPointerHotLink、58-esriPointerPencil、59-esriPointerHand、60-esriPointerPageZoomIn、61-esriPointerPageZoomOut、62-esriPointerPagePan、63-esriPointerPagePanning、99-esriPointerCustom。因为本身 MousePointer 属性就是 esri Controls MousePointer 的枚举类型，所以只需要在触发事件代码中添加如下代码：

```
//改变鼠标形状
axMapControl1.MousePointer = esriControlsMousePointer.esriPointerPan;
//执行平移操作
axMapControl1.Pan();
```

旋转地图时只需在触发事件中设置 MapControl 的 Rotation 属性即可,该属性以度为单位。

(3) 鼠标拉框放大。拉框放大时鼠标形状也会发生变化,参照以上 MousePointer 属性修改。拉框放大的过程实际上是利用 MapControl 的 TrackRectangle 方法获取新的显示范围,并将该显示范围赋值给当前 MapControl 的 Extent 属性,然后刷新 MapControl 以更新显示内容,参考代码:

```
//改变鼠标形状
axMapControl1.Extent = pmapcon2.TrackRectangle();
//改变鼠标形状
axMapControl1.Refresh();
```

自主练习:对照 OMD 尝试修改 MapControl 的其他属性并调用 MapControl 的其他方法。

4. PageLayoutControl 控件

PageLayoutControl 控件对应于 ArcMap 中的布局视图,并封装了 PageLayout 对象,提供了布局视图中控制地图元素的属性和方法。与 MapControl 控件相似,使用 ArcMap 创作的地图文档可以加载到 PageLayoutControl 中,用于编辑和打印地图。同时 PageLayoutControl 控件可以操作各种元素对象,如数据框、比例尺、指北针、地图标题、描述性文本和符号图例等,该部分元素操作内容将在后续地图整饰部分给出。

自主练习:按照 3 中的练习使用 PageLayoutControl 加载地图文档,对照 OMD 尝试修改 PageLayoutControl 的其他属性、调用 PageLayoutControl 的其他方法,理解 PageLayoutControl 与 MapControl 的区别和联系。

5. TOCControl 控件

TOCControl 控件用于实时交互式显示树状"伙伴控件"的地图、图层和符号体系内容,对应主要用于组织和管理数据框及其包含图层的状态与属性的内容表(table of content,TOC)。TOCControl 控件属于框架控件,要与一个且只能是一个"伙伴控件"协同工作,"伙伴控件"可以是 MapControl、PageLayoutControl、SceneControl 或 GlobeControl。例如,当"伙伴控件"是 MapControl 时,从该 MapControl 中删除了一个图层,则该图层也会从 TOCControl 中删除;同样地,在 TOCControl 中取消了某个图层的 Visibility 复选框,则该图层在 MapControl 中不再可见。在程序设计阶段可以通过 TOCControl 属性页的 Buddy 选项设置 TOCControl 的伙伴控件。

在程序运行过程中 TOCControl 的当前活动视图和伙伴控件往往会发生改变。例如,在数据视图和布局视图切换的时候,当前活动视图和伙伴控件会在 MapControl 和 PageLayoutControl 之间变换。此时可以通过编程的方式调用 TOCControl 的 SetActiveView 方法来设置当前活动视图,需要注意的是 SetActiveView 方法会默认设置 TOCControl 的 Buddy 属性为 None,因此在调用完 SetActiveView 方法之后需要调用 TOCControl 的 SetBuddyControl 方法来重新设置伙伴控件。

参考代码:

```
// axMapControl1.Object 返回对象名为 axMapControl1 的 MapControl 控件
IMapControl2 P_MapControl = axMapControl1.Object as IMapControl2;
//设置 TOC 的当前活动视图,该方法会导致 Buddy 的属性为 None
axTOCControl1.SetActiveView(P_MapControl.ActiveView);
```

//设置 TOC 的伙伴控件

  axTOCControl1.SetBuddyControl(P_MapControl);

  TOCControl 往往会通过鼠标对数据框及其包含的图层等进行快捷响应，而 TOCControl 树状显示"伙伴控件"的地图、图层和符号体系内容，这就需要 TOCControl 可以准确反馈点击的对象到底是数据框、图层还是图例。

  TOCControl 封装了 HitTest 方法来获取用户在 TOCControl 中点击的相关信息。HitTest 方法包含七个参数：用户点击的整型坐标 x、y；ref 型传递的 esriTOCControlItem 用于反馈点击对象类型；ref 型传递的 IbasicMap 用于反馈点击的地图对象；ref 型传递的 Ilayer 用于反馈点击的图层对象；ref 型传递的 object 用于反馈点击的图例组；ref 型传递的 object 用于反馈点击的图例的索引号。其中 esriTOCControlItem 为枚举类型，其值见表 2-1-2。通过 ref 传递参数解决了函数返回多个值的问题。此外还有 GetSelectItem 方法用于获得 TOCControl 被选择的项，SelectItem 方法用于设置 TOCControl 中的选择项。使用 HitTest 方法引用库中应当有 ESRI.ArcGIS.Geodatabase，可以不使用。

表 2-1-2　枚举类型 esriTOCControlItem 的值

枚举常量	值	描述
esriTOCControlItemNone	0	No item
esriTOCControlItemMap	1	The item is a map
esriTOCControlItemLayer	2	The item is a layer
esriTOCControlItemHeading	3	The item is a heading
esriTOCControlItemLegendClass	4	The item is a legend class

  自主练习：按照示例程序查看 TOC 反馈中的要素。对照 OMD 尝试修改 TOCControl 的其他属性、调用 TOCControl 的其他方法，如 EnableLayerDragDrop 属性，控制着图层是否可以通过拖放来改变图层顺序。

6. ToolbarControl 控件

  ToolbarControl 控件也是框架控件，要与一个"伙伴控件"协同工作。"伙伴控件"可以是 MapControl、PageLayoutControl、SceneControl 或 GlobeControl。它包含一组命令、工具、工具控件、工具菜单、工具面板，与伙伴控件配合使用。在程序设计阶段可以通过 ToolbarControl 属性页设置伙伴控件，也可以使用 ToolbarControl 的 SetBuddyControl 方法设置伙伴控件。

  使用钩子（hook）来联系命令对象与 MapControl 或 PageLayoutControl 等控件，并提供属性、方法、事件等。ToolbarControl 可以用于管理控件的外观，设置伙伴控件，添加、删除命令项，设置当前工具，定制工具。

  1）ToolbarControl 上的命令

  当一个命令对象驻留在 ToolbarControl 后，将调用 ICommand.OnCreate()，该方法传递一个 hook 参数给应用程序（hook 为 ToolbarControl 的伙伴控件）。首先测试该命令是否支持 hook 对象，如果不支持该 hook 对象，那么该命令无效；如果支持该 hook 对象，命令将存储该 hook 对象，以便以后使用。例如，打开地图文档命令是仅用于 MapControl 的。当 MapControl 作为 hook 参数被传递给打开地图文档命令的 OnCreate()时，该命令直接保存该 hook 参数以便后面使用。当 ToolbarControl 作为 hook 参数被传递给该打开地图文档命令的 OnCreate 事件时，该命令将检测该 ToolbarControl 的伙伴控件的类型，如果该伙伴控件为 MapControl，保存该 hook 参数以便后面使用；如果该伙伴控件为 GlobeControl，则该命令无效。

当用户点击驻留在 ToolbarControl 上的命令项时，将调用该命令的 ICommand.OnClick() 方法。依据命令项的类型，命令将使用 hook 参数访问来自于伙伴控件的对象，执行相应的功能。

命令项有以下三种类型。

（1）简单命令（Command）：当用户点击时，将调用该命令的 ICommand.OnClick()方法，执行相应的功能。

（2）工具（Tool）：需要用户与伙伴控件交互，才能完成需要的功能。当用户点击 ToolbarControl 上的某一工具时，该工具就成为该 ToolbarControl 的 CurrentTool，ToolbarControl 将其设置为伙伴控件 CurrentTool，并将接收来自于伙伴控件的鼠标、键盘事件。

（3）工具控件（ToolControl）：类似于下拉列表框，驻留在 ToolbarControl 上的一个小窗口，由 IToolControl.hWnd 提供窗口句柄。只能向 ToolbarControl 添加特定工具控件的一个实例。

设计时，可以通过 ToolbarControl 的属性对话框将命令项添加到 ToolbarControl，也可以通过编程将命令项添加到 ToolbarControl。使用 IToolbarControl.AddItem 方法有三种方式将命令项添加到 ToolbarControl 上，分别是 UID（使用 GUID）、ProgID 和 ICommand，三种方式的添加命令参考代码如下：

```
//通过 UID 添加命令
UID uID = new UIDClass();
uID.Value= "esriControls.ControlsAddDataCommand";
axToolbarControl1.AddItem(uID, -1, -1, false, 0,
esriCommandStyles.esriCommandStyleIconOnly);
//通过 ProgID 添加命令
String progID = "esriControls.ControlsOpenDocCommand";
axToolbarControl1.AddItem(progID, -1, -1, false, 0,
esriCommandStyles.esriCommandStyleIconOnly);
//通过 ICommand 添加命令
Icommand command = new ControlsMapFullExtentCommandClass();
axToolbarControl1.AddItem(command, -1, -1, false, 0,
esriCommandStyles.esriCommandStyleIconOnly);
```

其中 AddItem 方法的第 1 个参数为添加的命令对象，第 2 个参数为是否有子类型，第 3 个参数为该命令在工具条上的位置索引，第 4 个参数为是否有竖线标志新的命令组，第 5 个参数为新命令组与之前的命令组的间隔，需要与第 4 个参数一起使用，最后一个参数为 esriCommandStyles 枚举类型。esriCommandStyles 需要引用 SystemUI 库，其值见表 2-1-3。

表 2-1-3 枚举类型 esriCommandStyles 的值

枚举常量	值	描述
esriCommandStyleTextOnly	0	Display text only
esriCommandStyleIconOnly	1	Display icon only
esriCommandStyleIconAndText	2	Display icon and text
esriCommandStyleMenuBar	4	Display bar as main menu

默认情况下，ToolbarControl 每半秒钟自动更新其自身一次，以确保驻留在 ToolbarControl 上的每个工具条命令项的外观与其底层命令的 Enabled、Bitmap 和 Caption 等属性同步。可以通过改变 ToolbarControl 更新的频率来改变 ToolbarControl 的 UpdateInterval 属性。UpdateInterval 为 0 会停止任何自动发生的更新，开发人员必须编程调用 Update 方法以刷新每个工具条命令项的状态。在应用程序中首次调用 Update 方法时，ToolbarControl 会检测每个工具条命令项的 ICommand.OnCreate 方法是否已经被调用过。如果还没有调用过该方法，则该 ToolbarControl 作为钩子（hook）被自动传递给 ICommand.OnCreate 方法。

2）ToolbarItem

ToolbarItem 是驻留在 ToolbarControl 或菜单上的单个 Command、Tool、ToolControl、ToolbarMenu、ToolbarPalette 或 MultiItem 对象。从 OMD 中可以看出，ToolbarItem 以 0 个或多个的方式组成了 ToolbarControl，同时也以 0 个或多个的方式组成了 ToolbarMenu 和 ToolbarPalette。ToolbarItem 是普通类，因此获得 ToolbarItem 的方法是使用 ToolbarControl 的 GetItem 方法。

ToolbarItem 的属性决定工具条命令项的外观。例如，工具条命令项是否在其左侧有表示开始一个命令组（Group）的垂直线；命令项的样式（Style）是否有位图、标题或两者都有。ToolbarItem 的 Command 和 Menu 属性返回工具条命令项代表的实际命令或菜单。ToolbarItem 的 Type 属性返回 ToolbarItem 的类型，是枚举类型 esriToolbarItemType，其值可参考表 2-1-4。

表 2-1-4 枚举类型 esriToolbarItemType 的值

枚举常量	值	描述
esriToolbarItemUnknown	0	The item is not initialised
esriToolbarItemCommand	1	The item is a single click command implementing ICommand
esriToolbarItemTool	2	The item is a tool implementing ICommand and ITool
esriToolbarItemToolControl	3	The item is a tool control
esriToolbarItemMenu	4	The item is a ToolbarMenu
esriToolbarItemPalette	5	The item is a ToolbarPalette
esriToolbarItemMultiItem	6	The item is a ToolbarMultiItem

3）ToolbarMenu

ToolbarMenu（工具菜单）提供了菜单项的实现，其上可以驻留命令（Command）、工具（Tool）、工具控件（ToolControl）、动态菜单（MultiItem）和工具条面板（ToolbarPalette）。ArcGIS 10 之前版本，ToolbarMenu 只能驻留单击命令，不能驻留其他对象。ToolbarMenu 可以驻留在 ToolbarControl、ToolbarMenu 或作为右键菜单出现。以工具菜单驻留在 ToolbarControl 为例，其实现代码如下：

　　//创建工具菜单，定义工具菜单接口并实例化
　　IToolbarMenu2 toolbarMenu = new ToolbarMenuClass();
　　//为工具菜单添加命令
　　toolbarMenu.AddItem("esriControls.ControlsSelectTool", 0, -1, false, esriCommandStyles.esriCommandStyleIconAndText);
　　//将工具菜单添加到 ToolbarControl 上
　　axToolbarControl2.AddMenuItem(toolbarMenu, -1, true, 0);

　　4）ToolbarPalette

ToolbarPalette(工具面板)提供了面板项的实现，其上可以驻留命令和工具，但不能驻留 ToolControl、ToolbarMenu 及 MultiItem 对象。ToolbarPalette 可以驻留在 ToolbarControl 上或作为弹出面板出现。ToolbarPalette 驻留在 ToolbarControl 上时只显示已经选择命令的图标，其实现方法与工具菜单驻留在 ToolbarControl 类似。

　　5）CommandPool

每个 ToolbarControl、ToolbarMenu 和 ToolbarPalette 都有一个命令池（CommandPool），用于管理其使用的命令对象的集合。一般来说，开发人员不会与命令池进行交互。当通过 ToolbarControl 属性页或编程将命令添加到 ToolbarControl 中时，该命令自动添加到命令池中。命令对象要么作为唯一识别该命令的一个 UID 对象（使用 GUID）、要么作为命令对象的一个现有实例被添加到命令池中。

如果命令对象的一个现有实例被添加，并且该命令没有一个 UID，则命令池中可以有同一命令的多个实例存在。如果提供了一个 UID 对象，命令池可以确定该命令是否已经存在于命令池中，而且如果存在的话就可以重用该命令之前的实例。命令池通过追踪是否已经调用过命令的 OnCreate 方法来完成这个工作。如果已经调用过 OnCreate 方法，则将重用该命令并增加其使用次数（UsageCount）。

例如，如果向某个 ToolbarControl 中添加两次"ZoomIn"工具并提供 UID，则当 ToolbarControl 上的一个"ZoomIn"工具被选择并显示"按下"时，另一个"ZoomIn"工具也会显示"按下"状态，因为它们使用同一个命令对象。当应用程序包含多个 ToolbarControl 或工具条菜单时，开发人员应确保每个 ToolbarControl 和工具条菜单使用相同的命令池，以保证在应用程序中只创建了命令的一个实例。参考代码如下：

　　//设置同一命令池
　　IcommandPool commandpool = new CommandPoolClass();
　　axToolbarControl1.CommandPool = commandpool;
　　axToolbarControl2.CommandPool = commandpool;

　　6）OperationStack

ToolbarControl 有一个操作栈（OperationStack），用于管理"撤销（undo）"和"重做（redo）"功能。由每个工具条命令项的底层命令将操作添加到操作栈中，以便可以根据需要将操作前翻或后滚。例如，由于误操作删除了某个地理要素，可以点击 ToolbarControl 上的"撤销（undo）"命令撤销该操作。需要指出的是，当前活动视图（ActiveView）中的范围变化保存在 IActiveView

的 ExtentStack 中，而不是在 OperationStack 中。

命令是否可以利用操作栈取决于该命令的实现。典型情况下，开发人员为应用程序创建一个单个的控件操作栈（ControlsOperationStack），并将其设置给每个 ToolbarControl。撤销和重做命令可以添加到使用了操作栈的 ToolbarControl 上。当应用程序包含多个 ToolbarControl 或工具条菜单时，开发人员应确保每个 ToolbarControl 和工具条菜单使用相同的操作栈，以保证在应用程序中只创建了命令的一个实例，其实现方法与 CommandPool 类似。

7）程序运行期间自定义工具条

ToolbarControl 的命令可以在程序运行期间进行自定义，需要将其 Customize 属性设置为 true，即 ToolbarControl 处于定制模式。此时开发人员可以编程启动非模态定制对话框（CustomizeDialog），利用定制对话框列出的所有的控件命令及任何自定义命令、工具集和菜单来重新安排、删除和添加命令项，以及改变这些命令项的外观。

实现自定义工具条的关键步骤有三步：定义 Customize 属性为 true；启动 CustomizeDialog 并将拟自定义的工具条句柄作为参数传入；设置 CustomizeDialog 的双击对象。

参照代码如下：

```
//定义工具条 Customize 属性为 true
axToolbarControl1.Customize = true;
//声明 ICustomizeDialog 接口并实例化
IcustomizeDialog customzedialog = new CustomizeDialogClass();
//启动 CustomizeDialog 对话框，并传入 axToolbarControl1.hWnd
customzedialog.StartDialog(axToolbarControl1.hWnd);
//设置双击相应对象
customzedialog.SetDoubleClickDestination(axToolbarControl1);
```

然而我们往往喜欢使用 VS.NET 平台的 CheckBox 一键设置以上三个内容。为此我们需要设置两个事件代理，核心代码如下：

```
//声明事件代理
ICustomizeDialogEvents_OnStartDialogEventHandler startDialogE;
ICustomizeDialogEvents_OnCloseDialogEventHandler closeDialogE;
//添加事件代理
startDialogE = new
ICustomizeDialogEvents_OnStartDialogEventHandler(OnStartDialogHandler);
pCustomizeDialogEvent.OnStartDialog += startDialogE;
closeDialogE = new
ICustomizeDialogEvents_OnCloseDialogEventHandler(OnCloseDialogHandler);
pCustomizeDialogEvent.OnCloseDialog += closeDialogE;
//定义代理事件打开时的触发内容
private void OnStartDialogHandler()
{
    axToolbarControl1.Customize = true;
}
```

```
//定义代理事件关闭时的触发内容
private void OnCloseDialogHandler()
{
    axToolbarControl1.Customize = false;
    checkBox1.Checked = false;
}
```

思考题：在使用 ToolbarControl 属性页进行命令项添加的时候，Map Navigation 目录下有 Full Extent 命令，Globe 目录下也有 Full Extent 命令，Scene 目录下也有 Full Extent 命令，试分析这三个目录下同一名称的 Full Extent 命令的区别与联系。

自主练习：尝试将一个工具条的命令池直接赋值给另一个工具条并分析结果。对照 OMD 尝试修改 ToolbarControl 相关对象的其他属性、调用 ToolbarControl 相关对象的其他方法。

7. SymbologyControl

SymbologyControl 用于显示服务器样式文件（*.ServerStyle）和自定义符号系统的内容。如果安装了 ArcGIS for Desktop，则还可以显示样式文件（*.Style）的内容。SymbologyControl 使最终用户能够选择可应用于部分应用程序的单个符号，如图层的渲染器或元素的符号。

实验 2-2　地图文档及相关对象

在 ArcGIS 中，地图文档是包含一个或多个地图、一个页面布局及关联的图层、表格、图表和报告的文件，其文件的扩展名为 mxd。地图文档是保存和分发地图及其他制图信息或数据的最常用方式。我们使用 ArcMap 往往是从地图文档开始的。本实验利用一定的实例介绍地图文档相关对象，如果要深入理解还需要课下对照 OMD 和帮助进行学习。

（1）实验目的：从地图文档出发，理解地图文档、地图、图层等概念，并与控件建立对应关系。

（2）相关实验：实验 2-1　ArcGIS Engine 控件的使用。

（3）实验数据：ArcGIS Engine 自带的示例数据或本教材系列实验数据。

（4）实验环境：Visual Studio 2012、ArcGIS Engine 10.2 和 C#语言；本实验程序内大部分变量会作为类成员变量进行声明和实例化。

（5）实验内容：地图文档相关程序设计、地图对象相关程序设计、图层对象相关程序设计等。

1. 地图文档

许多常见的开发任务都是从地图文档开始的，因此了解如何访问和使用地图文档是 ArcObjects 开发的重要组成部分。在 ArcGIS 中可以通过多个不同组件访问和打开地图文档。选择的组件通常取决于具体情况和所使用的 ArcGIS 产品。对于 ArcGIS for Desktop 开发人员，访问和修改地图文档最常用的方法是通过 MxDocument 类。对于 ArcGIS Engine 开发人员，访问和修改地图文档的方法是通过 MapControl、PageLayoutControl、MapDocument 或 MapReader 类。

MxDocument 是 ArcMap 应用程序中加载的地图文档。通常，在使用 ArcMap 应用程序中通过钩子（hook）获取对 MxDocument 的引用。MxDocument 在 ArcMap 应用程序的进程空间内运行，创建 MxDocument 类实例会创建 ArcMap 进程，所以应当在 ArcMap 进程中访问 MxDocument 类。MxDocument 一般用于 Add-in 编程，本实验不做过多介绍，也不设计相应实验。

　　MapDocument 类提供了读取和修改地图文档的常用属性功能，是地图文档的简化形式。MapDocument 对批处理操作、低级文档访问及简单地图文档的修改非常有用。例如，要更新许多地图文档文件的内容，可以编写一个应用程序来遍历文件，使用 MapDocument 打开并更新内容，然后保存文档。使用地图文档相关元素时需要引用库中含有 ESRI.ArcGIS.Carto，其命名空间与库名相同，故在程序中需要写 using ESRI.ArcGIS.Carto。

　　1）新建地图文档

　　打开 ArcMap 的时候实际上都会让我们选择空文档或模板。因此，新建地图文档的一种思路是在硬盘上新建一个空 mxd 文件，将该空 mxd 文件中的地图赋值给 MapControl 控件。该方法的缺点是必须新建一个空的 mxd 文件。参考代码如下：

```
//在具体位置生成空地图文档
m_MapDocument.New("E:\\new.mxd");
//将空地图文档赋值给当前地图控件，其中 0 是索引号
m_mapControl.Map = m_MapDocument.Map[0];
```

　　2）打开地图文档

　　打开地图文档的思路是使用打开文档浏览器找到需要打开的地图文档，在打开地图文档前需先判断地图文档是否受保护及是否存在。将该地图文档的路径和名称一起赋值给 MapDocument 的 Open 方法，进一步将 MapDocument 的 Map 属性赋值给 MapControl 的 Map 属性，参考以下代码：

```
//使用地图文档的打开方法
m_MapDocument.Open(m_mapDocumentPathName, string.Empty);
//将 MapDocument 的 Map 属性赋值给 MapControl 的 Map 属性
m_mapControl.Map = m_MapDocument.Map[0];
```

　　3）保存地图文档

　　保存地图文档的思路是判断现在打开的地图文档是否是只读，如果不是只读则调用 MapDocument 的 Save 方法。参考以下代码：

```
//保存地图文档
m_MapDocument.Save(m_MapDocument.UsesRelativePaths, true);
```

　　需要提醒的是，此时的保存需要读取 MapDocument 的信息，因此 MapDocument 不能为空。例如，使用 ToolbarControl 添加的打开文档命令时，MapDocument 的内容为空，此时该方法会出错。

　　4）另存为地图文档

　　另存为地图文档需要有一个交互的过程，为此可以使用 VS.NET 平台的 OpenFileDialog 方法，通过交互获得将要保存地图文档的路径和文件名，然后调用 MapDocument 的 SaveAs 方法。参考以下代码：

```
//保存文档对话框
SaveFileDialog SFD = new SaveFileDialog();
```

```
        SFD.Filter = "地图文档 (*.mxd)|*.mxd";
        SFD.Title = "打开地图文档";
        if (SFD.ShowDialog() = = DialogResult.OK)
        {
            m_mapDocumentPathName = SFD.FileName;
            //另存为地图文档
            m_MapDocument.SaveAs(m_mapDocumentPathName,false, true);
        }
```

自主练习：对照 OMD 和帮助查看地图文档相关其他属性和方法。

2. Map 对象

地图（在 ArcMap 用户界面中称为数据框）是用于在 ArcGIS 中显示和组织地理信息的主要对象，为此，Map 作为图层集合进行维护。Map 对象还具有对地图中所有图层进行操作的属性，即空间参考、地图比例、标注引擎及操纵地图图层的方法。此外，Map 还具有辅助分析和导航地图全局属性。

Map 对象是开发任务的主要切入点，它不仅管理着数据，还是良好的可视化工具。Map 对象的典型任务包括添加新图层、平移显示、更改视图范围（缩放功能）、更改空间参考及获取当前选定的要素和元素等。与大多数类一样，在应用程序中 Map 可以被创建为对象使用。但是，更多情况是通过其他更高级别对象的地图文档获得对现有地图的引用。实例化新的 Map 对象会自动创建它依赖的 ScreenDisplay 对象和新的 CompositeGraphicsLayer。

1) IMap 接口

IMap 接口是许多地图操作任务的起点，它可以添加、删除和访问各种包含要素图层和图形图层的数据；也可以将地图环绕对象（如图例、比例尺等）与地图关联；还可以访问地图的各种属性，包括感兴趣区域（area of interest, AOI）、当前地图单位和空间参考、选择要素和访问 Map 对象的当前选择集。

2) IGraphicsContainer 接口

IGraphicsContainer 接口用于管理 Map 中的元素对象，该接口中的 AddElement 和 Delete Elements 方法分别用于地图对象添加和删除元素；其 LocateElements 和 LocateElementsByEnvelope 方法分别用于以点为圆心按照一定的范围选择元素和鼠标拖拽矩形区域的方式选择元素。

3) IActiveView 接口

主程序一般是由视图（ActiveView）控制，以 ArcMap 为例，其有两个视图对象：Map（数据视图）和 PageLayout（布局视图）。每个视图都有一个 ScreenDisplay 对象，用于执行绘图操作。ScreenDisplay 对象还使客户端可以创建任意数量的缓存。缓存是表示应用程序窗口的屏幕外位图。图形不是直接绘制到屏幕上，而是绘制到缓存中，然后在屏幕上绘制缓存。当应用程序的窗口被遮挡并需要重绘时，它是通过缓存而不是数据库完成的。通过这种方式，缓存可以提高绘图性能（位图渲染比从数据库中读取和显示数据更快）。

IActiveView 接口用于管理地图的显示，提供了视图相关的属性和方法。它还提供对地图 ScreenDisplay 的指针。该接口可以显示或隐藏标尺与滚动条、刷新地图和局部刷新地图等。进行局部刷新地图的方法是 PartialRefresh。在进行要素选择操作时，需要调用 PartialRefresh

两次,即要素选择前调用 PartialRefresh 一次,要素选择后调用一次。表 2-2-1 给出了部分刷新时地图和布局的响应。

表 2-2-1　部分刷新时地图和布局的响应

刷新语句	地图响应	布局响应
esriViewBackground	Map grids	Page/snap grid
esriViewGeography	Layers	Unused
*esriViewGeoSelection	Feature selection	Unused
esriViewGraphics	Labels/graphics	Graphics
esriViewGraphicSelection	Graphic selection	Element selection
esriViewForeground	Unused	Snap guides

4）IMapBookmarks 接口

空间书签是用于保存特定视图的范围,以方便快速定位、查看和浏览的工具。ImapBookmarks 是管理 Map 中的空间书签对象的接口,其主要属性方法为 AddBookmark(IspatialBookmark bookmark)、RemoveAllBookmarks、RemoveBookmark。进行空间书签记录的是 IspatialBookmark,有两个类可以实现该接口：用感兴趣区域定义空间书签的 AOIBookmark 类和用选择要素的空间覆盖范围定义书签的 FeatureBookmark 类。

5）设置全景显示

全景显示的思路是获得 MapControl 当前视图的 IActiveView 接口,将 IActiveView 的 FullExtent 属性赋值给 IActiveView 的 Extent 属性,然后刷新地图。参考以下代码：

  m_ActiveView.Extent = m_ActiveView.FullExtent;

  m_ActiveView.Refresh();

6）MapControl 与 PageLayoutControl 初步联动

数据视图的内容主要用于数据编辑、查询、分析和可视化,很多情况下需要在布局视图中进行地图整饰和输出,因此需要数据视图和布局视图中的数据改变是实时互动的。ArcGIS Engine 的 MapControl 和 PageLayoutControl 两个控件本身并不能自动实时互动,只能通过编程复制的方法来实时传递两个控件的内容。

在窗体设计阶段仿照 ArcMap 添加一个 VS.NET 自带的容器 TabControl,其中 tabPage1 改名为数据视图用于放置 MapControl,tabPage2 改名为布局视图用于放置 PageLayoutControl。当 TabControl 进行切换时会触发 SelectIndexChanged 事件,在该事件中进行判断,当 SelectedIndex 是 0 的时候是设置 TOC 和 Toolbar 活动视图和伙伴控件地图控件；当 Selected Index 是 1 的时候 TOC 和 Toolbar 是布局控件,并进行地图复制。其中 IObjectCopy 和 Object CopyClass 需引用 System 库,参考代码如下：

  if (tabControl1.SelectedIndex == 0)

  {

    //设置 TOC 当前活动视图和伙伴控件

    axTOCControl1.SetActiveView(m_mapControl.ActiveView);

    axTOCControl1.SetBuddyControl(m_mapControl.Object);

```csharp
        //设置 ToolbarControl 伙伴控件
        axToolbarControl1.SetBuddyControl(m_mapControl.Object);
}
else if (tabControl1.SelectedIndex == 1)
{
        //设置 TOC 当前活动视图和伙伴控件
        axTOCControl1.SetActiveView(m_pageLayoutControl.ActiveView);
        axTOCControl1.SetBuddyControl(m_pageLayoutControl.Object);
        //设置 ToolbarControl 伙伴控件
        axToolbarControl1.SetBuddyControl(axPageLayoutControl1.Object);
        //进行地图复制，获得布局视图
        IActiveView pActiveView = axPageLayoutControl1.ActiveView.FocusMap as IActiveView;
        //布局视图与地图视图同步
        pActiveView.ScreenDisplay.DisplayTransformation.VisibleBounds = m_mapControl.Extent;
        pActiveView.Refresh();
        //进行地图复制，将地图控件的 Map 复制到布局视图的 FocusMap
        IObjectCopy pObjectcopy = new ObjectCopyClass();
        object copymap = pObjectcopy.Copy(m_mapControl.Map);
        object overwritemap = axPageLayoutControl1.ActiveView.FocusMap;
        pObjectcopy.Overwrite(copymap, ref overwritemap);
}
```

7）构建要素选择集并清空

构建要素选择集的思路是使用 MapControl 的 TrackRectangle 方法获得选择范围存储于 IEnvelope。IEnvelope 需要引用库 ESRI.ArcGIS.Geometry，该库的命名空间与库名相同。使用 IMap 的 SelectByShape 获得要素选择集，使用 IActiveView 的 PartialRefresh 刷新显示。由于该操作需要有鼠标按下交互操作，宜将代码写到 MapControl 的 OnMouseDown 内，并设置触发区别（如设置一个 m_flag）。参考代码如下：

```csharp
//获得选择范围，需要 Geometry 库
IEnvelope pEnv = m_mapControl.TrackRectangle();
//进行选择
m_map.SelectByShape(pEnv, null, false);
//部分刷新获得选择集
m_ActiveView.PartialRefresh(esriViewDrawPhase.esriViewGeoSelection, null, null);
```

清空选择集的思路是：使用 IMap 的 ClearSelection 方法清空，并使用 IActiveView 的 PartialRefresh 刷新显示。参考代码如下：

```csharp
m_map.ClearSelection();
m_ActiveView.Refresh();
```

8）构建元素选择集

元素选择集的构建需要使用 IGraphicsContainer 接口以点为圆心按照一定的范围选择元素的 LocateElements 方法，或者鼠标拖拽矩形区域的方式选择元素的 LocateElementsByEnvelope 方法。这两个方法只对 Element 有效，返回的是枚举类型的 IEnumElement。参考代码如下：

IEnumElement pEnumEle = pGraphicsC.LocateElementsByEnvelope(pEnv);

9）空间书签

空间书签的实现至少需要三步：①记录空间书签；②创建空间书签；③定位空间书签。对于记录空间书签可以采用 ToolbarMenu + IMultiItem + Form、ComboBox + Form 或 TreeView + Form 等形式；对于已创建感兴趣区域的空间书签需要使用 IAOIBookMark 和 IMapBookmarks；对于定位空间书签，相对于第一种记录方式则需要 IMapBookmarks、ISpatialBookMark 和 IMultiItem。以下采用第一种方式实现空间书签，首先实现一个 Form 用于交互。

我们在 ToolbarControl 上添加一个 ToolbarMenu 用于记录空间书签，参考代码如下：

```
m_ToolbarMenu = new ToolbarMenuClass();
m_ToolbarMenu.Caption = "空间书签";
m_ToolbarMenu.AddItem(new CreatBookmark(), 0, -1, false,
esriCommandStyles.esriCommandStyleTextOnly);
m_ToolbarMenu.AddMultiItem(new SpatialBookmarks(), -1, false,
esriCommandStyles.esriCommandStyleTextOnly);
axToolbarControl1.AddItem(m_ToolbarMenu, 0, -1, true, 0,
esriCommandStyles.esriCommandStyleMenuBar);
```

使用 IAOIBookMark 记录 IActiveView 的 Extent，使用 IMapBookmarks 记录 IAOIBookMark，参考代码如下：

```
//Get the focus map
IActiveView activeView = (IActiveView)m_HookHelper.FocusMap;
//Create a new bookmark
IAOIBookmark bookmark = new AOIBookmarkClass();
//Set the location to the current extent of the focus map
bookmark.Location = activeView.Extent;
//Set the bookmark name
bookmark.Name = sName;
//Get the bookmark collection of the focus map
IMapBookmarks mapBookmarks = (IMapBookmarks)m_HookHelper.FocusMap;
//Add the bookmark to the bookmarks collection
mapBookmarks.AddBookmark(bookmark);
```

定位空间书签时，一方面需要在 ToolbarMenu 上找到空间书签的名称，另一方面需要在 IMultiItem 上找到具体的空间书签，参考代码如下：

```
public void OnItemClick(int index)
```

```
    {
        //Get the bookmarks of the focus map
        IMapBookmarks mapBookmarks = (IMapBookmarks)m_HookHelper.FocusMap;
        //Get bookmarks enumerator
        IEnumSpatialBookmark enumSpatialBookmarks = mapBookmarks.Bookmarks;
        enumSpatialBookmarks.Reset();
        //Loop through the bookmarks to get bookmark to zoom to
        ISpatialBookmark spatialBookmark = enumSpatialBookmarks.Next();
        int bookmarkCount = 0;
        while (spatialBookmark != null)
        {
            //Get the correct bookmark
            if (bookmarkCount = = index)
            {
                //Zoom to the bookmark
                spatialBookmark.ZoomTo(m_HookHelper.FocusMap);
                //Refresh the map
                m_HookHelper.ActiveView.Refresh();
            }
            bookmarkCount = bookmarkCount + 1;
            spatialBookmark = enumSpatialBookmarks.Next();
        }
    }
```

自主练习：对照 OMD 和帮助查看地图相关其他属性和方法，并尝试调用。提前学习本节实验涉及的 Geometry、Display、ADF、System 和 SystemUI 库相关内容。

3. Layer 对象

Map 对象可以装载地理数据，这些数据是以图层的形式组织到地图对象中的，图层（Layer）对象在地图上显示地理信息。ArcGIS 中也可以在一个要素类上新建一个图层文件，即 lyr 文件，这个文件只是获取了地理数据的硬盘位置，而没有实际数据。

图层（Layer）对象不存储实际的地理数据，它仅引用地理数据库、coverage、Shapefile、栅格数据集、网格（grids）等地理数据，然后定义如何将这些地理数据可视化。某些图层不涉及地理数据，例如，GroupLayer 用于包含其他图层，CompositeGraphicsLayer 用于存储图形。

Map 对象中显示地图时会设置空间参照系（SpatialReference），当第一个图层添加到 Map 中时，Map 对象的空间参照系就自动设置为这个图层的空间参照系，假设后面还有新的图层加入该 Map 对象，无论后面新加入的图层是否含有空间参照系都会使用 Map 对象已经设置的空间参照系进行显示。

所有的图层对象都实现了 ILayer 接口，该接口定义了所有图层的公共属性和方法，如 Name 属性用于设置或获得图层名字；MaximunScale 和 MinimunScale 用于显示或设置图层可以出现的最大和最小比例尺，在该范围之外，图层上的数据不显示；ShowTips 属性用于指示当鼠标放在图层中某个要素上的时候，是否会出现提示信息；TipText 属性用于确定图层提示信息；SpatialReference 属性用于设置图层的空间参照系；AreaOfInterest（只读）属性用于获得图层的覆盖范围；Cached 属性用于管理图层是否拥有自己的缓存；Valid 属性用于管理图层是否有效；Visible 属性用于管理图层是否可视；SupportedDrawPhases 属性用于获取图层支持哪些层面绘制（esriDPGeography-1、esriDPAnnotation-2、esriDPSelection-4）；当 SupportedDrawPhases =3 时，图层支持地理要素和注记两方面的绘制。

ILayer2 接口新增了可读可写的 AreaOfInterest 属性和只读的 ScaleRangeReadOnly 属性。AreaOfInterest 属性可以用来设置图层的覆盖范围；ScaleRangeReadOnly 属性可以获得 MaximunScale 和 MinimunScale 是否只读，确定图层是否可以设置 MaximunScale 和 MinimunScale 属性。从 OMD 上看 ILayer2 接口并没有继承 ILayer 接口，在程序设计时需要使用查询接口进行两个接口之间的转换。下面是 ArcObjects 10.2 给出的图层种类及其功能描述，见表 2-2-2。

表 2-2-2 图层种类及其功能描述

图层名	描述
BasemapLayer	底图图层
CadAnnotationLayer	CAD 注记图层
CadFeatureLayer	CAD 要素图层
CadLayer	CAD 图层
CompositeGraphicsLayer	符合图形图层
CoverageAnnotationLayer	Coverage 注记图层
DimensionLayer	Geodatabase 尺寸标注要素图层
DummyGraduatedMarkerLayer	在样式库中显示分级标记图例项
DummyLayer	在样式库中显示分级标记图例项
FDOGraphicsLayer	显示地理数据库注记功能
FDOGraphicsSublayer	在地理数据库注记要素类的子集中显示注记要素
FeatureLayer	矢量要素图层
GdbRasterCatalogLayer	显示存储在地理数据库中的栅格目录数据
GraphicsSubLayer	CompositeGraphicsLayer 的图形子层
GroupLayer	一组行为类似于单个图层的图层
IMSMapLayer	显示 IMS 数据
IMSSubFeatureLayer	IMSMapLayer 的功能子层
IMSSubLayer	IMSMapLayer 的子层
MapServerBasicSublayer	basicMapServerLayer 子层
MapServerFindSublayer	具有查找功能的 MapServerLayer 子图层
MapServerIdentifySublayer	具有标识功能的 MapServerLayer 子图层
MapServerLayer	显示 ArcGIS Map Server 数据
MapServerQuerySublayer	具有查找和识别功能的 MapServerLayer 子图层
MosaicLayer	显示镶嵌数据集
NetworkLayer	显示网络数据集

续表

图层名	描述
RasterBasemapLayer	通过优化的绘图路径显示栅格以提高性能
RasterCatalogLayer	显示栅格目录数据
RasterLayer	显示栅格数据
TerrainLayer	显示地形数据
TinLayer	显示 TIN 数据
TopologyLayer	将地理数据库拓扑显示为图层
WMSGroupLayer	WMSMapLayer 的子层组
WMSLayer	WMSMapLayer 的子层
WMSMapLayer	将 WMS 服务显示为图层

1）设置所有图层可见

设置所有图层可见的思路是定义一个 ILayer 接口对象，制作一个循环遍历获得 Map 中的所有图层。设置图层的可见属性，参照以下代码：

```
ILayer PLayer = null;
for (int i = 0; i < m_map.LayerCount; i++)
{
    PLayer = m_map.get_Layer(i);
    PLayer.Visible = true;
}
```

2）访问获得地图和图层名称

访问获得地图和图层名称的思路有两种，一种是和以上 1）中代码类似的循环遍历获得图层，然后打印图层名称；另一种是使用 Map 的 get_Layers 方法获得枚举类型的 IEnumLayer，然后遍历 IEnumLayer，打印图层名称。第二种方式的核心代码如下：

```
IEnumLayer pEnumLayer = pMap.get_Layers(null, true);
pEnumLayer.Reset();
ILayer pLayer = pEnumLayer.Next();
while (pLayer != null)
{
    MessageBox.Show(pLayer.Name);
    pLayer = pEnumLayer.Next();
}
```

自主练习：对照 OMD 和帮助查看图层相关其他属性和方法，并尝试调用。提前学习本节实验涉及的 Geometry、Display、ADF、System 和 SystemUI 库相关内容。

实验 2-3 几何对象与空间参考

地理信息系统研究和处理的对象是地理空间数据，而地理空间数据的表达需要几何对象和空间参考的支持。

（1）实验目的：明确几何对象类型的特点，理解 ArcObjects 空间参考系统，熟练使用几何对象和空间参考的属性和方法。

（2）相关实验：实验 2-2 地图文档及相关对象。

（3）实验数据：ArcGIS Engine 自带的示例数据或本教材系列实验数据。

（4）实验环境：Visual Studio 2012、ArcGIS Engine 10.2 和 C#语言。

（5）实验内容：几何对象绘制；空间参考设置。

1. 几何对象

地理数据库和图形元素系统使用矢量数据格式来定义要素和图形的形状。Geometry 库提供了几何符号系统的定义和操作，并将几何对象分为两个层次：高级几何对象（用于定义要素的几何形状）和构件几何对象（用于构建高级几何对象）。

1）高级几何对象

高级几何对象：Point、Multipoint、Polyline、Polygon 和 MultiPatch 拥有几何拓扑约束。当满足所有拓扑约束时，几何形状被认为是简单的；当违反拓扑约束时，或者不知道是否满足拓扑约束时，几何形状被认为是非简单的。ITopologicalOperator、IPolygonN 和 IPolylineN 接口提供测试和实施简单化操作。除了 X 和 Y 坐标之外，高级几何对象的每个顶点都可以具有其他属性，称为顶点属性，如 Z、M 和 pointID。

高级几何对象支持经典的集合理论操作，用于生成新的几何对象。集合运算包括：并（Union）、按结构并（ConstructUnion）、交（Intersection）、差（Difference）和对称差分（Symmetric Difference，也即异或运算），这些集合运算通过 ITopologicalOperator 接口实现，计算会创建新几何对象作为结果。

高级几何对象也支持 IRelationalOperator 接口，该接口可以对一对几何对象执行各种测试，如是否相离（Disjoint）、是否相接（Touch）、是否包含（Contain）等，返回 Bool 型结果。这些接口的三维版本（如 IRelationalOperator3D）在对 Z-aware 几何体执行操作时会考虑 Z 属性。在确定结果时，ITopologicalOperator 和 IRelationalOperator 都使用与输入几何对象一致的空间参考。

2）构件几何对象

构件几何对象：Paths、Rings、Segments、TriangleStrips、TriangleFans 和 Triangles。

路径（Paths）用于构建多段线（Polyline），多段线包含路径。

环（Rings）用于构建多边形（Polygon），多边形包含环。

三角形带（TriangleStrips）、三角形扇（TriangleFans）及三角形（Triangles）用于构建多面（MultiPatch），多面包含三角形带、三角形扇、三角形和环。

路径和环是由线段（Segment）相连的顶点序列。线段是参数化的函数，用于定义连接顶点的几何形状。线段的类型包括圆弧（CircularArc）、直线（Line）、椭圆弧（EllipticArc）和贝塞尔曲线（BezierCurve）四种类型。

3）Envelope

Envelope，用于描述其他几何对象的空间范围，它是一个矩形（仅考虑 X、Y 坐标时）。它覆盖了几何对象的最小坐标和最大坐标、Z 值和 M 值的变化范围。

4）GeometryBag

GeometryBag，是任何类型的几何体对象的集合，实际上 GeometryBag 是一个可以容纳任何类型几何对象的容器，可以同时容纳多种不同类型的几何对象。

5）Polyline

多段线（Polyline）对象是相连或不相连的路径对象的有序集合，它可以分别是单个路径（Path）、多个不相连的路径（Path）和多个相连路径（Path）的集合。

路径（Path）是连续线段（Segment）对象的集合，除了路径的第一个和最后一个 Segment 外，每一个 Segment 的起始点都是前一个 Segment 的终止点，即路径对象中的 Segment 不能出现分离。

路径（Path）可以是任意数目的 Line、CircularArc、EllipticArc 和 BezierCurve 的组合。一个或多个路径对象组成一个 Polyline 对象。

Segment 对象是一个继承于 Curve 的抽象类。Curve 也是抽象类，它定义了起点（FromPoint）和终点（ToPoint）。ISegment 是 Segment 对象的主要接口，它提供加密（densify）线段、分割线段、获取线段的曲率等方法。Segment 对象还支持 ISegmentM、ISegmentZ 和 ISegmentID 三个接口来分别获得和设置 Segment 对象的 M 值、Z 值和 ID 号。

Ring、Path、Polyline 和 Polygon 等都可以由 Segment 对象集合创建。其中 Ring 和 Path 支持 ISegmentColection 接口，Polyline 和 Polygon 支持 IGeometryColection 接口。通过 ISegmentColection 接口的 AddSegment、RemoveSegment、AddSegmentCollection 和 SegmentCount 等方法可以将 Segment 集合构建成直线（Line）、圆弧（CircularArc）、椭圆弧（EllipticArc）和贝塞尔曲线（BezierCurve）。

（1）直线（Line）。直线（Line）是最简单的线段，由起点和终点决定一条直线段，是一维集合对象。直线（Line）通常用于构造 Ring、Path、Polyline 和 Polygon 等对象。

（2）圆弧（CircularArc）。圆弧（CircularArc）对象是圆的一部分，如果使用圆弧（CircularArc）来表示一个整圆，需设置 CentralAngle 为 2π，同时设置起点和终点是同一个点。几何对象中圆弧（CircularArc）是椭圆弧（EllipticArc）的特例。

ICircularArc 是圆弧（CircularArc）的主要接口，可以通过该接口获得 FromAngle、ToAngle、CentralAngle、CentralPoint、ChordHeight 和 Radius 属性。ICircularArc 的 Complement 方法可以将没有闭合的圆弧闭合，返回一个整圆对象。

（3）椭圆弧（EllipticArc）。椭圆弧（EllipticArc）对象是椭圆的一部分。椭圆是通过一个长轴、一个短轴、中心点和旋转角度值来确定的几何对象。如果旋转角度为 0，则椭圆对象的两个轴分别与 X、Y 轴重合，另外椭圆弧（EllipticArc）也可以使用 FromAngle 属性和 ToAngle 属性确定形状。

（4）贝塞尔曲线（BezierCurve）。贝塞尔曲线（BezierCurve）是由四个控制点定义的。贝塞尔曲线与控制点 0 和控制点 1 构成的线段、控制点 2 和控制点 4 构成的线段相切，这四个控制点产生了一条平滑的曲线。

IBezierCurve 接口定义了 PutCoord 和 PutCoords 方法，其中 PutCoord 只能用于修改贝塞尔曲线上的点；PutCoords 可以用于生成新的贝塞尔曲线，但是因为它是一个点一个点进行设置，所以需要一次性同时设置四个控制点。如果有一个已经设置好的控制点（IPoint）数组，可以考虑使用 IBezierCurveGEN 接口，该接口的 PutCoords 方法可以将控制点（IPoint）数组作为输入参数。

（5）IGeometryCollection。IGeometryCollection 接口提供了构建 Multipoint、Polyline、

Polygon、MultiPatch 和 GeometryBag 的方法。对于 Multipoints、Triangles、TriangleFans 和 TriangleStrips，其连接是基于 Point；对于 Polylines，其连接是基于 Path；对于 Polygon，其连接是基于 Ring；对于 MultiPatches，其连接是基于 Triangles、TriangleFans、TriangleStrips 和 Ring；对于 GeometryBags，其连接可以基于任意的 IGeometry 对象。

6）Polylgon

多边形（Polygon）对象是一个有序环（Ring）对象的集合，这些环可以是一个或者多个。多边形对象通常可以用于描述具有面积的多边形离散矢量对象。

环是一种闭合的路径（Path），因此起始点具有相同的坐标。环有内外环之分，通过顺时针围绕顶点旋转 Polygon 在右手边为外环，Polygon 在左手边为内环。环具有以下三个特征：

（1）环包含一系列首尾相连的同方向 Segment 对象。

（2）环是封闭的，起点终点一致。

（3）环不能自相交。

7）MultiPatch

MultiPatch 描述具有纹理的三维几何对象表面。三维几何对象还可以存储顶点法线（vertex normal）、顶点 ID、顶点的 M 值及基于部分的属性（part-level attributes）。MultiPatch 可以包含材质（指定颜色、纹理、透明度信息）和纹理坐标（指定每部分上面纹理的放置）。ArcScene 提供的样式库中包含 MultiPatch 模型，如三维的建筑物、树、车辆、街道设施及其他主题。

MultiPatch 由一到多个 Triangles、TriangleFans、TriangleStrips 和 Rings 组成。可通过导入多种不同文件格式（3D Studio Max 的 3ds 文件，Sketchup 的 skp 文件）的数据来创建 MultiPatch，也可以通过编程以多种不同的方式来创建 MultiPatch。

（1）不带纹理、法线和部分属性的 MultiPatch 可以这样定义：创建部分（parts），然后创建 MultiPatch，再使用 IGeometryCollection 接口将创建的部分添加到 MultiPatch 中。

（2）带纹理、法线和部分属性的 MultiPatch 需要使用 GeneralMultiPatchCreator（在拥有 3D Analyst 许可的情况下）来创建。IGeneralMultiPatchInfo 接口可以获得已经存在的 MultiPatch 的法线和材质信息。

8）GeometryEnvironment

GeometryEnvironment 是一个单实例对象（singleton object），因此只有在第 1 次使用 new 字段实例化 GeometryEnvironmentClass 时，才会创建一个新的 GeometryEnvironment 对象。第 1 次以后使用 new 字段实例化 GeometryEnvironmentClass，指挥返回第 1 次创建的 GeometryEnvironment 对象的引用。GeometryEnvironment 用于控制创建几何对象的行为，提供了根据不同的输入创建几何对象的方法，即设置或获取全局变量。其主要接口为 IGeometryBridge 和 IGeometryBridge2 接口。

9）交互绘制点

交互绘制点的思路是使用 ScreenDisplay 对象的 DisplayTransformation 方法获得鼠标点击的点，使用 IGraphicsContainer 的 AddElement 方法将点添加到地图上，最后刷新显示。其中 IMarkerElement 需要 Carto 库支持，ISimpleMarker 需要 Display 库支持并引用。

参考示例代码如下：

　　//交互获得点

```
IPoint pPt = m_ActiveView.ScreenDisplay.DisplayTransformation.ToMapPoint(e.x, e.y);
IMarkerElement pMarkerElement = new MarkerElementClass();
ISimpleMarkerSymbol pMarkerSymbol = new SimpleMarkerSymbolClass();
pMarkerSymbol.Size = 4;
pMarkerSymbol.Style = esriSimpleMarkerStyle.esriSMSCross;
IElement pElement = pMarkerElement as IElement;
pElement.Geometry = pPt;
pMarkerElement.Symbol = pMarkerSymbol;
IGraphicsContainer pGraphicsContainer = m_map as IGraphicsContainer;
pGraphicsContainer.AddElement(pMarkerElement as IElement, 0);
m_ActiveView.PartialRefresh(esriViewDrawPhase.esriViewGraphics, null, null);
```

10）交互绘制线、多边形、矩形

交互绘制线、多边形和矩形的思路是使用 IRubberBand 接口分别定义 RubberLineClass 对象、RubberPolygonClass 对象和 RubberRectangularPolygonClass 对象。使用 IRubberBand 接口的 TrackNew 方法分别获得线、多边形和矩形，再使用 IGraphicsContainer 的 AddElement 方法将点添加到地图上，最后刷新显示。其中 IRubberBand 和 IScreenDisplay 需要 Display 库支持并引用，用于获得相应的线、多边形、矩形和圆。以下为绘制线、多边形、矩形的示例代码。

（1）绘制线。
```
IScreenDisplay pScreenDisplay = m_ActiveView.ScreenDisplay;
//定义橡胶线
IRubberBand pRubberPolyline = new RubberLineClass();
ISimpleLineSymbol pLineSymbol = new SimpleLineSymbolClass();
//交互获得多段线
IPolyline pPolyline = pRubberPolyline.TrackNew(pScreenDisplay, (ISymbol)pLineSymbol) as IPolyline;
```

（2）绘制多边形。
```
IScreenDisplay pScreenDisplay = m_ActiveView.ScreenDisplay;
//定义橡胶多边形
IRubberBand pRubberPolygon = new RubberPolygonClass();
ISimpleFillSymbol pFillSymbol = new SimpleFillSymbolClass();
//交互获得多边型
IPolygon pPolygon = pRubberPolygon.TrackNew(pScreenDisplay, (ISymbol)pFillSymbol) as IPolygon;
```

（3）绘制矩形。
```
IScreenDisplay pScreenDisplay = m_ActiveView.ScreenDisplay;
//定义橡胶矩形
IRubberBand pRubberPolygon = new RubberRectangularPolygonClass();
ISimpleFillSymbol pFillSymbol = new SimpleFillSymbolClass();
```

//交互获得矩形

IPolygon pPolygon = pRubberPolygon.TrackNew(pScreenDisplay, (ISymbol)pFill Symbol) as IPolygon;

11）交互绘制圆

交互绘制圆的思路是使用 IRubberBand 接口定义 RubberCircleClass 对象，再使用 IRubberBand 接口的 TrackNew 方法获得圆。IGraphicsContainer 的 AddElement 方法可以添加 ILineElement 和 IFillShapeElement，但无法添加简单的圆弧。为此需要将圆弧构造成 Polyline 或者 Ploygon 才可以使用 IGraphicsContainer 的 AddElement 方法。使用 ISegmentCollection 的 AddSegment 方法将通过查询接口从 IGeometry 的圆弧转换到 ISegment 的线段添加到 Ploygon 中，然后可以使用 IGraphicsContainer 的 AddElement 方法将点添加到地图上，最后刷新显示。其中从 IGeometry 的圆弧转换到 ISegment 线段的过程为 IGeometry 到 IConstructCircularArc、IConstructCircularArc 到 ISegment；IRubberBand 和 IScreenDisplay 需要 Display 库支持并引用。

使用 ScreenDisplay 对象的 DisplayTransformation 方法获得鼠标点击的点，使用 IGraphicsContainer 的 AddElement 方法将点添加到地图上，最后刷新显示。其中 IRubberBand 和 IScreenDisplay 需要 Display 库支持并引用。核心代码参考如下：

```
IScreenDisplay pScreenDisplay = m_ActiveView.ScreenDisplay;
//定义橡胶圆
IRubberBand pRubberCircle = new RubberCircleClass();
ISimpleFillSymbol pFillSymbol = new SimpleFillSymbolClass();
//交互获得圆
IGeometry pCircle = pRubberCircle.TrackNew(pScreenDisplay, (ISymbol)pFillSymbol) as IGeometry;
IConstructCircularArc pConstructArc = pCircle as IConstructCircularArc;
IPolygon pPolygon = new PolygonClass();
ISegmentCollection pSegmentCollection = pPolygon as ISegmentCollection;
ISegment pSegment = pConstructArc as ISegment;
object missing = Type.Missing;
//通过 ISegmentCollection 的 AddSegment 方法将线段添加到 Ploygon
pSegmentCollection.AddSegment(pSegment, ref missing, ref missing);
```

12）构建绘制多段线

构建绘制多段线的思路是使用 WKSPoint 对象生成一系列的点，再通过 IPointCollection 构建多段线对象，再使用 IGeometryBridge2 的 AddWKSPoints 方法将 WKSPoint 对象生成一系列的点添加到 IPointCollection 构建的多段线对象，然后通过查询接口将 IPointCollection 转换到 IPolyline 接口获得 Polyline 对象，最后使用 IGraphicsContainer 的 AddElement 方法将点添加到地图上，刷新显示。其中 WKSPoint 对象需要 System 库支持，核心代码参考如下：

```
//构建一系列点
WKSPoint[] wksPoints = new WKSPoint[100];
for (int i = 0; i < wksPoints.Length; i++)
{
```

```
            wksPoints[i].X = i * 10;
            wksPoints[i].Y = i * 10;
}
//定义多段线的点集
IPointCollection4 pointCollection = new PolylineClass();
IGeometryBridge2 geometryBridge = new GeometryEnvironmentClass();
//一系列点添加到多段线的点集
geometryBridge.AddWKSPoints(pointCollection, ref wksPoints);
IPolyline polyline = pointCollection as IPolyline;
```

13）绘制 MultiPoint

绘制 Multipoint 可以参照"构建绘制多段线"中的方式，通过 IPointCollection 构建 Multipoint 对象，再使用 IGeometryBridge2 的 AddWKSPoints 方法将 WKSPoint 对象生成一系列的点添加到 IPointCollection 构建的 Multipoint 对象。IGraphicsContainer 的 AddElement 方法不支持 Multipoint 元素的添加，可以使用 IScreenDisplay 的 DrawMultipoint 方法。核心代码参考如下：

```
//构建一系列点
WKSPoint[] wksPoints = new WKSPoint[50];
for (int i = 0; i < wksPoints.Length; i++)
{
            wksPoints[i].X = (i+1) +50*i;
            wksPoints[i].Y = (i+1) +50*i;
}
//定义 Multipoint
IPointCollection4 pointCollection = new MultipointClass();
IGeometryBridge2 geometryBridge = new GeometryEnvironmentClass();
//一系列点添加到 Multipoint
geometryBridge.AddWKSPoints(pointCollection, ref wksPoints);
IMultipoint multipoint = pointCollection as IMultipoint;
ISimpleMarkerSymbol pMarkerSymbol = new SimpleMarkerSymbolClass();
pMarkerSymbol.Style = esriSimpleMarkerStyle.esriSMSCircle;
pMarkerSymbol.Size = 15;
IRgbColor rgb = new RgbColorClass();
rgb.Green = 255;
pMarkerSymbol.Color=rgb;
rgb.Green = 0;
rgb.Red = 255;
pMarkerSymbol.Outline = true;
pMarkerSymbol.OutlineSize = 2;
pMarkerSymbol.OutlineColor = rgb;
```

```csharp
//使用 IScreenDisplay 绘制 Multipoint
IScreenDisplay screenDisplay = m_ActiveView.ScreenDisplay;
screenDisplay.StartDrawing(screenDisplay.hDC,(short)esriScreenCache.esriNoScreenCache);
screenDisplay.SetSymbol((ISymbol)pMarkerSymbol);
screenDisplay.DrawMultipoint(multipoint as IGeometry);
screenDisplay.FinishDrawing();
```

14）绘制贝塞尔曲线

贝塞尔曲线的绘制需要定义四个控制点，可以参照从 BezierCurve 到 Segment，再从 Segment 到 Path，再从 Path 到 Polyline 的方式构建。可以在 IGraphicsContainer 上绘制 Line Element，核心代码参考如下：

```csharp
IScreenDisplay pScreenDisplay = m_ActiveView.ScreenDisplay;
ISimpleLineSymbol pLineSymbol = new SimpleLineSymbolClass();
//定义贝塞尔曲线，使用 IBezierCurveGEN 的 PutCoords 添加点
IBezierCurveGEN bezier = new BezierCurveClass();
//定义 IPoint 数组
IPoint[] point=new IPoint[4];
//每一个点都要实例化
point[0] = new PointClass();
point[1] = new PointClass();
point[2] = new PointClass();
point[3] = new PointClass();
point[0].PutCoords(50,200);
point[1].PutCoords(200, 400);
point[2].PutCoords(50, 50);
point[3].PutCoords(400, 50);
bezier.PutCoords(ref point);
object Missing = Type.Missing;
//通过 ISegmentCollection 定义 Path，添加 Segment（通过 as）
ISegmentCollection segmentCollection1 = new PathClass();
segmentCollection1.AddSegment(bezier as ISegment, ref Missing, ref Missing);
//通过 IGeometryCollection 的 AddGeometry 方法获得 Polyline
IGeometryCollection geometryCollection = new PolylineClass();
geometryCollection.AddGeometry(segmentCollection1 as IGeometry, ref Missing, ref Missing);
pLineSymbol.Style = esriSimpleLineStyle.esriSLSSolid;
ILineElement pPolylineEle = new LineElementClass();
pPolylineEle.Symbol = pLineSymbol;
IElement pEle = pPolylineEle as IElement;
pEle.Geometry = geometryCollection as IPolyline;
```

```
IGraphicsContainer pGraphicsContainer = m_map as IGraphicsContainer;
pGraphicsContainer.AddElement(pEle, 0);
m_ActiveView.PartialRefresh(esriViewDrawPhase.esriViewGraphics, null, null);
```

自主练习：对照 OMD 尝试绘制其他几何对象，调用其他属性和方法。

2. 空间参考

空间参考（spatial reference）是空间数据的参照基准，其能够将空间数据定位到相应的位置，为地图中的每一点提供准确的坐标。Geodatabase 中新建一个要素数据集或一个单独的要素类都必须设置它们的空间参考。空间参考包括以下属性：坐标系、分辨率与空间域、容差，其中坐标系有地理坐标系、投影坐标系和竖直坐标系三种类型；分辨率是指坐标精细程度，即坐标值小数点后的有效位数；空间域是指要素类 X、Y、Z 和 M 的取值范围；容差是用于聚集操作时设置坐标之间的最小距离，如拓扑验证、缓冲区生成和多边形叠加等。

ArcObjects 包含大量预定义的空间参考系及空间参考系的构建块（building blocks）。每个预定义的对象由一个工厂代码（a factory code）识别。工厂代码由 esriSR 开头的枚举集合定义，一般使用枚举宏（enumeration macro）而不是整数值来生成预定义的对象。

1）状态栏

ArcGIS 程序中空间参考的信息一般是在数据的属性页或者状态栏中显示，为此可以使用 VS.Net 平台提供的 StatusStrip 进行空间参考信息的部分信息显示。我们使用 StatusStrip 的 StatusLabel 来显示信息，当鼠标在 MapControl 中移动时触发信息显示，其中 e.mapX 和 e.mapY 显示了空间参考系下的坐标值，MapUnits 显示了该空间参考的坐标单位。

参考示例代码：

```
toolStripStatusLabel1.Text = string.Format("{0}, {1}   {2}", e.mapX.ToString ("#######.##"), e.mapY.ToString("#######.##"), axMapControl1.MapUnits.ToString(). Substring(4));
```

2）地理坐标系

地理坐标系（geographic coordinate system）包含坐标系的名称、角度单位、大地基准（datum，包含椭球体）及本初子午线（prime meridian）。其中名称是地理坐标系的名称；角度单位可以选择弧度、角度等并可以定义角度单位的尺度；大地基准是定义旋转椭球面来模拟地球，需要用长半轴和短半轴或扁率来定义；本初子午线用于确定 0 经度的起算点。地理坐标系对象通过 IGeographicCoordinateSystem2 接口提供的属性、方法可以访问、创建地理坐标系的相关对象。

地理坐标系的定义往往比较困难，需要大量的大地测量工作支撑，一般情况下，很少会碰到自定义的地理坐标系的情况，但如果需要自定义地理坐标系，则可以调用 IGeographicCoordinateSystemEdit 接口的 Define 或 DefineEx 方法。

定制地理坐标系可以参考以下代码：

```
ISpatialReferenceFactory3 spatialReferenceFactory = new SpatialReferenceEnvironmentClass();
//定义 Datum
IDatum datum = spatialReferenceFactory.CreateDatum((int)esriSRDatum Type.esriSRDatum_OSGB1936);
//定义 PrimeMeridian
```

IPrimeMeridian primeMeridian=spatialReferenceFactory.CreatePrimeMeridian((int) esriSRPrimeMType.esriSRPrimeM_Greenwich);

//定义 Unit

IUnit unit = spatialReferenceFactory.CreateUnit((int)esriSRUnitType.esriSRUnit_ Degree);

IGeographicCoordinateSystemEdit geographicCoordinateSystemEdit = new GeographicCoordinateSystemClass();

object name = "UserDefined Geographic Coordinate System";

object alias = "UserDefined GCS";

object abbreviation = "UserDefined";

object remarks = "User Defined Geographic Coordinate System based on OSGB1936";

object usage = "Suitable for the UK";

object datumObject = datum as object;

object primeMeridianObject = primeMeridian as object;

object unitObject = unit as object;

//调用 Define 函数

geographicCoordinateSystemEdit.Define(ref name, ref alias, ref abbreviation, ref remarks, ref usage, ref datumObject, ref primeMeridianObject, ref unitObject);

IGeographicCoordinateSystem userDefinedGeographicCoordinateSystem = geographicCoordinateSystemEdit as IGeographicCoordinateSystem;

m_map.SpatialReference = userDefinedGeographicCoordinateSystem;

m_ActiveView.Refresh();

MessageBox.Show("已将当前坐标系统转换为自定义 GeographicCoordinateSystem！");

3）投影坐标系

投影坐标系（projected coordinate system）是以地理坐标系为基础的。投影是从曲面到平面的一个过程，其中曲面即地理坐标系的旋转椭球面，平面即最后需要的 X、Y 平面。因此投影坐标系的组成部分有：投影名称（projection）、线性单位（linear unit）和地理坐标系（geographic coordinate system）。其中投影决定了从曲面变换到平面的方式，线性单位决定了最后呈现的坐标单位。

投影坐标系支持 ISpatialReference2 和 ISpatialReferenceFactory。当定义定制的投影坐标系时，可以使用多种 esriSR 开头的枚举集合中预定义的对象。通过 IProjectedCoordinateSystem5 接口提供的属性、方法可以访问、创建投影坐标系相关对象。IProjectedCoordinateSystemEdit 接口的 Define 方法可用于定义定制的投影坐标系。

定制投影坐标系可以参照以下代码：

ISpatialReferenceFactory2 spatialReferenceFactory = new SpatialReferenceEnvironmentClass();

//定义投影名称和投影方式

IProjectionGEN projection = spatialReferenceFactory.CreateProjection((int)esriSRProjectionType.esriSRProjection_Sinusoidal) as IProjectionGEN;

//定义地理坐标系

IGeographicCoordinateSystem geographicCoordinateSystem = spatialReferenceFactory. CreateGeographicCoordinateSystem((int)esriSRGeoCSType.esriSRGeoCS_WGS1984);
//定义线性坐标长度
ILinearUnit unit = spatialReferenceFactory.CreateUnit((int)esriSRUnitType. esriSRUnit_Meter) as ILinearUnit;
IParameter[] parameters = projection.GetDefaultParameters();
IProjectedCoordinateSystemEdit projectedCoordinateSystemEdit = new Projected Coordinate SystemClass();
object name = "Newfoundland"; object alias = "NF_LAB";
object abbreviation = "NF";
object remarks = "Most Eastern Province in Canada";
object usage = "When making maps of Newfoundland";
object geographicCoordinateSystemObject = geographicCoordinateSystem as object;
object unitObject = unit as object;
object projectionObject = projection as object;
object parametersObject = parameters as object;
projectedCoordinateSystemEdit.Define(ref name, ref alias, ref abbreviation, ref remarks, ref usage, ref geographicCoordinateSystemObject, ref unitObject, ref projectionObject, ref parametersObject);
IProjectedCoordinateSystem userDefinedProjectedCoordinateSystem = projected Coordinate SystemEdit as IProjectedCoordinateSystem;
m_map.SpatialReference = userDefinedProjectedCoordinateSystem;
m_ActiveView.Refresh();
MessageBox.Show("自定义 ProjectedCoordinateSystem 完成！");

4）竖直坐标系

竖直坐标系（vertical coordinate system）或垂直坐标系用来定义竖直方向的高程基础。由于地球形状的不均一、质量的不均一，地球的大地水准面与我们定义的旋转椭球面往往不重合，而我们平时测量的高程都是以重力来找到铅锤方向的高程，即正高或正常高系统。旋转椭球面是一个数学上的连续曲面，大地水准面是一个物理上的连续曲面，为此需要定义一个竖直方向的基准面来确定高程的表征。

竖直坐标系的定义可参考以下代码：

ISpatialReferenceFactory3 spatialReferenceFactory = new SpatialReferenceEnvironment Class();
//定义竖直基准面
IVerticalDatum verticalDatum = spatialReferenceFactory.CreateVerticalDatum((int) esri SRVerticalDatumType.esriSRVertDatum_Taranaki);
IHVDatum hvDatum = verticalDatum as IHVDatum;
//定义竖直方向的线性单位
ILinearUnit linearUnit = spatialReferenceFactory.CreateUnit((int) esriSRUnitType. esri

SRUnit_Meter) as ILinearUnit;
　　IVerticalCoordinateSystemEdit verticalCoordinateSystemEdit = new VerticalCoordinateSystemClass();
　　object name = "New VCoordinateSystem";
　　object alias = "VCoordinateSystem alias";
　　object abbreviation = "abbr";
　　object remarks = "Test for options";
　　object usage = "New Zealand";
　　object hvDatumObject = hvDatum as object;
　　object unitObject = linearUnit as object;
　　object verticalShift = 40 as object;
　　object positiveDirection = -1 as object;
　　verticalCoordinateSystemEdit.Define(ref name, ref alias, ref abbreviation, ref remarks, ref usage, ref hvDatumObject, ref unitObject, ref verticalShift, ref positiveDirection);
　　IVerticalCoordinateSystem verticalCoordinateSystem = verticalCoordinateSystemEdit as IVerticalCoordinateSystem;
　　m_map.SpatialReference = verticalCoordinateSystem as ISpatialReference;
　　m_ActiveView.Refresh();
　　MessageBox.Show("自定义 verticalCoordinateSystem 完成！");

5）地理变换

由于地球的不均衡性和大地测量学的发展，旋转椭球面有很多定义，不同国家建立的基准也不一致。基准转换的需求往往很多，ArcGIS 定义的地理变换是一种数学运算，它是将点的坐标从一个地理坐标系变换到另一个地理坐标系的基准转换。地理变换的参数有名称、两个地理坐标系（源、目标地理坐标系）、变换的方法或类型、方法所需的参数。

地理变换的代码可以参考：

　　Type factoryType = Type.GetTypeFromProgID("esriGeometry.SpatialReference Environment");
　　System.Object obj = Activator.CreateInstance(factoryType);
　　ISpatialReferenceFactory2 spatialReferenceFactory2 = obj as ISpatialReferenceFactory2;
　　//Ordnance Survey Great Britain 1936(OSGB 1936)
　　//定义地理转换方法，这种情况是已知转换参数的形式
　　IGeoTransformation geoTransformation = spatialReferenceFactory2.CreateGeoTransformation((int)esriSRGeoTransformationType.esriSRGeoTransformation_OSGB1936_To_WGS1984_1) as IGeoTransformation;
　　ISpatialReference fromSpatialReference;
　　ISpatialReference toSpatialReference;
　　geoTransformation.GetSpatialReferences(out fromSpatialReference, out toSpatialReference);
　　System.Windows.Forms.MessageBox.Show("转换前坐标系："+geoTransformation.Name);

System.Windows.Forms.MessageBox.Show("转换后坐标系：" + fromSpatialReference.Name + ", " + toSpatialReference.Name);

自主练习：对照 OMD 和当今数据生产需求，设计实用的坐标系统定义和坐标转换软件。

实验 2-4 参数传递与鹰眼地图

程序实现的过程中，参数传递是再常见不过的了，而 ArcGIS Engine 的控件的加入，使得参数的传递更为复杂，因为经常会遇到在子窗体中调用主窗体或子类调用主类的控件的情况。

（1）实验目的：通过学习使用 ArcGIS Engine 的参数传递基本方法，构建鹰眼地图。
（2）相关实验：实验 2-1 ArcGIS Engine 控件的使用、实验 2-2 几何对象与空间参考"。
（3）实验数据：ArcGIS Engine 自带的示例数据或本教材系列实验数据。
（4）实验环境：Visual Studio 2012、ArcGIS Engine 10.2 和 C#语言。
（5）实验内容：参数传递方式；鹰眼地图制作。

1. 参数传递方式

VS.Net 平台本身提供了很多参数传递方式，如静态变量、ref 传递、Object 传递、委托传递、函数传递等。这些不同的参数传递方式对应不同的使用情况，而 ArcGIS Engine 中的控件如 MapControl、PageLayoutControl、SceneControl 和 GlobeControl 在很多情况下也需要作为参数进行传递，如实验 2-5 即将介绍的命令封装，需要将以上控件传递到命令中进行操作，这就是一种在子类中调用主类情况。本实验第 2 部分介绍的鹰眼地图，不仅在子窗体中显示主窗体中的地图全景，还可以在子窗体中对主窗体进行操作，如子窗体变换主窗体的显示范围等。这种情况往往需要钩子（hook）传递。

这些都需要在子类中调用主类中的 Engine 控件的情况，针对 MapControl 和 PageLayoutControl 等控件需要参数传递的情况，Controls 库提供了 HookHelper 类进行参数传递，提供在子类中调用主类控件的情况，它的实现是以一种钩子的形式，将主窗体中的 MapControl 和 PageLayoutControl 控件钩到子类中，进行操作时的感觉就像是子类自己的控件一样。HookHelper 类具有 FocusMap 和 PageLayout 属性，但没有 SceneControl 和 GlobeControl 相关的对象元素，也就是 HookHelper 类只支持 MapControl 和 PageLayoutControl 控件的钩子传递。从 Controls 库中可以看到支持 SceneControl 和 GlobeControl 进行钩子传递的是 SceneHookHelper 类和 GlobeHookHelper 类。

以在子类中钩子 HookHelper 类对象调用主类中的 MapControl 对象为例,设主窗体为 Main_Form，子窗体为 Eagle_Form。则在主窗体中需要定义钩子和 Eagle_Form 窗体。参考代码如下：

IHookHelper m_hook;
Eagle_Form eagle;
在 Main_Form 主窗体的 Form_Load 方法中进行钩子实例化和 Eagle_Form 窗体实例化：
m_hook = new HookHelperClass();
m_hook.Hook = m_mapControl.Object;
eagle = new Eagle_Form(m_hook);
因此在 Eagle_Form 窗体中需要定义两个 IMapControl3 接口，一个是主窗体的 Map

Control，另一个是子窗体的 MapControl：

IMapControl3 m_frmmain_mapControl;

IMapControl3 m_frmeagle_mapControl;

另外还需要改写构造函数，使得 IHookHelper 作为一个参数可以传递到 Eagle_Form 窗体，为此当 Eagle_Form 窗体 Show 事件触发时，即把主窗体的 MapControl 钩到子窗体中。改写构造函数参考代码如下：

```
public Eagle_Form(IHookHelper hook_fm1_map)
{
    InitializeComponent();
    m_frmmain_mapControl = hook_fm1_map.Hook as IMapControl3;
    m_frmeagle_mapControl = axMapControl1.Object as IMapControl3;
}
```

自主练习：分别尝试静态变量、ref 传递、Object 传递、委托传递、函数传递和钩子传递。

2. 鹰眼地图

本实例采用 ArcGIS Engine 自带的示例数据，其位于 ..\ArcGIS\DeveloperKit10.2\Samples\Data 中，采用两个窗体承载两个视图控件来进行鹰眼地图的实现。鹰眼的实现需要两个视图控件：主视图和鹰眼视图，其中鹰眼视图控件显示一幅地图的全景并使用矩形框标出主视图详细显示的范围；主视图中显示鹰眼视图中标出范围内的详细数据，并且当主视图移动范围的时候鹰眼视图的矩形框要一起移动；另外如果鹰眼视图中的矩形框被拖动或者重绘，则主视图中显示的范围也应当与鹰眼视图中的矩形框内标出的范围一致。这就需要：

（1）主视图和鹰眼视图共享一幅地图。

（2）在鹰眼视图中绘制出矩形框，该矩形框的范围与主视图的 Extent 一致，当 Extent 发生改变则矩形框也一起改变。

（3）鹰眼视图可以操作主视图，这个可以借鉴以上钩子传递，为此当矩形框重绘或被移动时，主视图中的显示范围和数据也一起发生改变。

（4）在鹰眼视图下打开窗体时，钩子传递，但关闭时需要主窗体相关信息一起发生改变，否则会出现 Bug。本实验此处采用委托的方式进行参数传递。

本实验采用的是两个窗体承载两个 MapControl。

1）视图内容同步

视图内容的同步可以采用复制或者添加图层的方式，因为图层不修改源数据，只是对数据源进行读取、组织和显示。为此可以采用读取图层的方式，参考代码如下：

```
for (int i = 0; i < m_frmmain_mapControl.Map.LayerCount; i++)
{
    m_frmeagle_mapControl.Map.AddLayer(m_frmmain_mapControl.Map.get_Layer(i));
}
```

以上代码即将主窗体中的 MapControl 的 Map 中的所有图层读取出来，然后添加到鹰眼窗体的 MapControl 的 Map 中，然而会出现图 2-4-1 的情况，即只看到面状最底下的图层和一些标注。这种情况是因为我们按照以上代码进行复制的时候将图层的顺序倒了过来，鹰眼窗体获得的第一个图层放到了最下面，而这第一个图层恰恰是应该在最上面的，以此类推。为此需要将以上代码修改，让鹰眼窗体的图层顺序与主窗体的图层顺序一致，可以改 i++ 为 i−−以倒序的方式进行图层复制，参见以下代码：

```
for (int i = m_frmmain_mapControl.Map.LayerCount - 1; i >= 0; i--)
{
    m_frmeagle_mapControl.Map.AddLayer(m_frmmain_mapControl.Map.get_Layer(i));
}
```

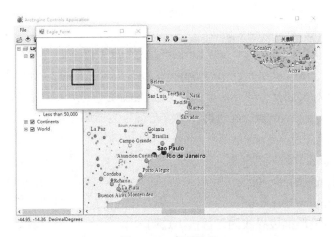

图 2-4-1 鹰眼异常

值得注意的是，即使我们正确地将图层复制过来，但有可能还是会出现图 2-4-1 的情况，这种情况有可能是窗体太小，导致控件太小，显示的内容也少，经过概括后或比例尺小于一定值之后数据不再显示。解决这种情况只需把视图控件拉大一些，能够显示内容即可。

2）矩形框绘制

有了实验 2-1～实验 2-3 的基础，绘制矩形框就比较简单了，但是需要考虑哪个窗体触发事件、哪个窗体响应事件并绘制矩形框。其思路是当主窗体触发地图可视范围改变事件时，调用鹰眼窗体的绘制矩形框函数，利用主窗体的地图可视新范围在鹰眼窗体中绘制新的矩形框。

参照代码如下：

```
axMapControl1_OnExtentUpdated(object sender, IMapControlEvents2_OnExtentUpdatedEvent e)
{
    if (m_eagle_flag == true)
    {
        IEnvelope penv = e.newEnvelope as IEnvelope;
        eagle.DrawRectangle(penv);
    }
}
```

绘制矩形框，不是绘制一个方形的环（Ring），而是绘制一个矩形的多边形（Polygon）。因为 Polygon 由两部分组成，一部分是边界线，另一部分是填充。将 Polygon 的填充部分设置为透明，边界线设置为红色，则红色矩形框即可轻松绘制。

参考代码如下：

```csharp
public void DrawRectangle(IEnvelope penv)
{
    //获得主窗体新的显示范围，是一个 Envelope
    IRectangleElement pRE = new RectangleElementClass();
    IElement pele = pRE as IElement;
    pele.Geometry = penv;
    IFillSymbol pfs = new SimpleFillSymbolClass();    //获得填充符号
    ILineSymbol pls = new SimpleLineSymbolClass();    //获得线符号
    IRgbColor prgb = new RgbColorClass();
    prgb.Transparency = 0;      //填充符号为透明
    pfs.Color = prgb;
    prgb.Red = 255;
    prgb.Blue = 0;
    prgb.Green = 0;
    //矩形框边界线为红色，不透明
    prgb.Transparency = 255;
    pls.Color = prgb;
    pls.Width = 3;
    pfs.Outline = pls;
    IFillShapeElement pfse = pele as IFillShapeElement;
    pfse.Symbol = pfs;
    //绘制矩形框
    IGraphicsContainer pGC = m_frmeagle_mapControl.Map as IGraphicsContainer;
    pGC.DeleteAllElements();
    pGC.AddElement(pfse as IElement, 0);
    IActiveView pac = pGC as IActiveView;
    pac.PartialRefresh(esriViewDrawPhase.esriViewGraphics, null, null);
}
```

3）鹰眼窗体与主窗体交互

鹰眼窗体与主窗体交互实际上就是拖动鹰眼窗体的矩形框时，主窗体的地图显示范围会变化；右键在鹰眼窗体绘制矩形框时，主窗体的地图显示范围会重置为矩形框选择区域。对于前者，由于拖动矩形框的时候，矩形框的大小并没有变化，也就是说主窗体的显示范围大小并没有变化，只是显示内容变化，即显示范围的中心点变化了；对于后者，矩形框变化导致主窗体的显示范围变化，为此可以将重绘的矩形框作为 Envelope 赋值给主窗体的范围（Extent）。

参考代码如下：
```
//鼠标按下事件
axMapControl1_OnMouseDown(object sender,IMapControlEvents2_OnMouseDownEvent e)
    {
        if (e.button == 1)
        {
            IPoint ppt = new PointClass();
            ppt.PutCoords(e.mapX, e.mapY);
            m_frmmain_mapControl.CenterAt(ppt);
            m_frmmain_mapControl.Refresh();
        }
        else if (e.button == 2)
        {
            m_frmmain_mapControl.Extent = m_frmeagle_mapControl.TrackRectangle();
            m_frmmain_mapControl.Refresh();
        }
    }
//鼠标移动事件
axMapControl1_OnMouseMove(object sender, IMapControlEvents2_OnMouseMoveEvent e)
    {
        if (e.button == 1)
        {
            IPoint ppt = new PointClass();
            ppt.PutCoords(e.mapX, e.mapY);
            m_frmmain_mapControl.CenterAt(ppt);
            m_frmmain_mapControl.Refresh();
        }
    }
```

自主练习：认真阅读代码，实际上新窗体作为鹰眼的载体，只在第一次比较好用，重复开关会出现问题，随书示例代码使用委托的方式修复了漏洞，试分析是否还有别的方法？另外，是否还可以定制其他个性化程序？

实验 2-5 命令封装与右键菜单

右键菜单提供了对视图、数据框和图层的快捷响应，为数据操作提供便利，右键菜单弹出后的按钮都是命令、工具、工具控件等。ArcGIS Engine 内置了很多实用的命令可以提高开发效率，另外封装命令也可以提高代码的重用粒度和可读性。

（1）实验目的：了解 ArcGIS Engine 的内置命令，学会自己封装命令，掌握右键菜单的制作方式。

（2）相关实验：实验 2-4 参数传递与鹰眼地图。

（3）实验数据：ArcGIS Engine 自带的示例数据或本教材系列实验数据。

（4）实验环境：Visual Studio 2012、ArcGIS Engine 10.2 和 C#语言。

（5）实验内容：ArcGIS Engine 的内置命令学习；命令封装；制作右键菜单。

1. 内置命令

ArcGIS Engine 内置了大量命令、工具条及工具、工具控件和菜单等，可以直接使用这些对象，利用内置命令、工具的名称、GUID（CLSID/ProgID）、子命令/子工具序号等信息可以查阅开发帮助文档。参考位置：Developing with ArcGIS→Learning ArcObjects→General ArcObjects references→Names and IDs，分为 ArcCatalog commands、ArcGlobe commands、ArcMap commands、ArcScene commands、ArcGIS Engine commands 和 Extensions 六类。

ESRI.ArcGIS.Controls 类库中提供了大量的命令、工具条及工具、工具控件和菜单等，在 GIS 应用开发中可以直接使用这些对象，只需要触发这些命令、工具的事件即可。对于命令型，往往只要通过 HookHelper、GlobeHookHelper 和 SceneHookHelper 对象将操作对象传递到命令内部，触发点击事件。参考代码如下：

```
//execute Open Document command
ICommand command = new ControlsOpenDocCommandClass();
command.OnCreate(m_mapControl.Object);
command.OnClick();
```

对于工具型，往往需要通过 HookHelper、GlobeHookHelper 和 SceneHookHelper 对象将操作对象传递到命令内部，然后设置操作对象的当前工具为该工具，之后等待触发鼠标交互事件。

参考代码如下：

```
ICommand pcom = new ControlsNewLineTool();
pcom.OnCreate(m_mapControl.Object);
m_mapControl.CurrentTool = pcom as ITool;
```

以上命令或工具等也可以添加到 ToolbarControl 上进行使用，采用 AddItem 方法添加。命令添加成功后会自动触发 OnCreate 方法，以钩子形式将伙伴控件传递，当点击命令时触发 OnClick 方法或设置 CurrentTool。

2. BaseCommand 基类与 ICommand 接口

在 ESRI.ArcGIS.ADF.Local 库中定义了一系列基本的命令、命令条、工具、工具条等基类用来继承，可以使用 ArcGIS Engine 自带的模板进行自定义命令的创建。以定义设置所有图层可见（TurnAllLayersOn）为例：首先在解决方案资源管理器中项目名称上鼠标右击，在弹出的右键菜单上选择"添加"→"新建项"，然后在 ArcGIS 选项卡下的 Extending ArcObjects 里选择 Base Command，最后在名称处填写 TurnAllLayersOn.cs。

点击确定后会弹出一个 ArcGIS New Item Wizard Options 对话框，里面会提示选择哪种命令类型，要注意此时一定要选择与将要处理的对象相匹配的命令类型。假设我们想要自定义命令设置 MapControl 的地图图层全部可见，但是我们选择的是第一项"ArcGlobe or Globe Control Command"，那么该命令将无法与 MapControl 进行参数传递响应，因为此时命令会使用 GlobeHookHelper 进行参数传递，而我们想要处理的 MapControl 是使用 HookHelper 进行参数传递的。

选择正确的命令类型，点击"OK"按钮，解决方案资源管理器中就会出现两个文件：TurnAllLayersOn.cs 和 TurnAllLayersOn.bmp，其中 bmp 文件就是命令的符号显示。假设我们把该命令添加到了 ToolbarControl 上，则 ToolbarControl 上显示的图标就是 TurnAllLayersOn.bmp。

打开 TurnAllLayersOn.cs 类可以看到，该类继承自 BaseCommand 类，Guid 和 ProgId 已经自动生成。在构造函数中可以看到两个 TODO，一个是改变继承自 BaseCommand 类的属性，其中 m_caption 是该命令进行显示时的名称、m_toolTip 是将该命令置于 ToolbarControl 上是弹出信息的标题、m_message 是将该命令置于 ToolbarControl 上时弹出信息的内容；另一个 TODO 提示可改变 bitmap name。

再往下可以看到一个 region。region 中有两个函数：一个是 OnCreate，负责参数传递和变量初始化；另一个是 OnClick，负责响应点击事件，这两个都是继承自 BaseCommand 类，这两个方法都采用了 override 的关键字。

由于我们自定义的命令有三种情况，即在按钮中触发、添加到 ToolbarControl 上、作为右键菜单弹出。这三种情况有两类参数传递，一类是传递 MapControl，另一类是传递 ToolbarControl，但是最终处理的是 ToolbarControl 的 Buddy。为了代码的完整性，我们需要修改 OnCreate 的代码以适应这两种情况。

参考代码如下：

```
// TODO: Add other initialization code
//get the MapControl from the hook in case the container is a ToolbarControl
if (m_hookHelper.Hook is IToolbarControl)
{
    m_mapControl = (IMapControl3)((IToolbarControl)m_hookHelper.Hook).Buddy;
}
//In case the container is MapControl
else if (m_hookHelper.Hook is IMapControl3)
{
    m_mapControl = (IMapControl3)m_hookHelper.Hook;
}
else
{
    MessageBox.Show("Active control must be MapControl!", "Warning", MessageBoxButtons.OK, MessageBoxIcon.Exclamation);
return;
}
m_map = m_mapControl.Map;
m_activeView = m_mapControl.ActiveView;
```

在 OnClick 中我们需要编写响应所有图层打开的代码，参考代码如下：
　　// TODO: Add TurnAllLayersOn.OnClick implementation
　　ILayer layer = null;
　　for (int i = 0; i < m_map.LayerCount; i++)
　　{
　　　　layer = m_map.get_Layer(i);
　　　　layer.Visible = true;
　　}
　　m_activeView.PartialRefresh(esriViewDrawPhase.esriViewGeography, null, m_activeView.Extent);

到此我们自定义所有图层打开（TurnAllLayersOn）的命令设置完毕。如果使用按钮调用自定义命令，可参照如下代码：
　　ICommand Pcom = new TurnAllLayersOn();
　　Pcom.OnCreate(m_mapControl.Object);
　　Pcom.OnClick();

实际上我们也可以不使用类模板，直接新建一个 VS.Net 平台下的一般类，让该类继承 BaseCommand 类即可，然后修改响应的属性和重载两个方法：OnCreate 和 OnClick。由于继承的是类，即使对继承基类的属性和方法一点也没有修改或 override 也不会报错。

在 ESRI.ArcGIS.SystemUI 库中定义了一系列基本的命令、命令条、工具、工具条等接口，这些接口也可以用来定义自定义命令。

以自定义所有图层关闭（TurnAllLayersOff）的命令为例：首先新建一个 VS.Net 平台下的一般类，将类名改为 TurnAllLayersOff.cs，让该类继承 ICommand 接口。ICommand 接口的属性和方法见图 2-5-1。与继承类不同的是，继承接口的话，需要实现所有的属性和方法，为此我们需要在新建的 TurnAllLayersOff.cs 中实现所有的 ICommand 的属性和方法，此时的方法前不能有 override 的关键字，因为不是修改，而是新建。另外为了实现 Bitmap 属性还需要寻找一个合适的 bmp 文件。设置属性的方法可参考如下设置"只读属性"的代码：

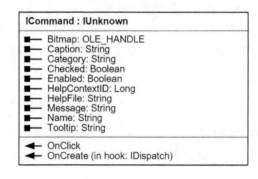

图 2-5-1　ICommand 接口

```
public string Caption
{
```

```
        get
        {
            return "Turn All Layers Off";
        }
    }
```

由于继承 ICommand 接口往往比较复杂，为此我们一般选择继承 BaseCommand 类。从 ICommand 接口的帮助文档中可以看到实现该接口的类有哪些，这些实现了 ICommand 接口的类都可以直接使用。

自主练习：尝试调用实现了 ICommand 接口的类。

3. BaseTool 类与 Itool 接口

BaseTool 类与 BaseCommand 类同属于 ESRI.ArcGIS.ADF.Local 库，但多了三个事件，分别是 OnMouseDown、OnMouseMove 和 OnMouseUp。这是为了应对绘制图像等工具命令时需要鼠标交互的情况。以画点（DrawPoint）为例，可以参照本章第 2 小节"BaseCommand 基类与 ICommand 接口"中利用项目模板新建 DrawPoint.cs。此时选择 Base Tool，选择正确的命令类型后点击"OK"发现解决方案资源管理器中出现了三个文件：DrawPoint.cs、DrawPoint.cur 和 DrawPoint.bmp。其中 DrawPoint.cur 是鼠标光标文件，因为我们设置的是鼠标点击处绘制点，所以需要将绘制点的代码写在 OnMouseDown 事件的响应区域。此时调用自定义工具时，不再是传完参数触发 OnClick，而是设置当前工具为自定义工具。

参考代码如下：

```
ICommand Pcom = new DrawPoint();
Pcom.OnCreate(m_mapControl.Object);
m_mapControl.CurrentTool = Pcom as ITool;
```

从 ESRI.ArcGIS.SystemUI 库的 OMD 中可以看到，如图 2-5-2 所示，ITool 接口并未定义 OnCreat 和 OnClick，而在 ITool 接口的帮助文档中可以看到很多类继承了该接口，如果细心一点会发现这些类同时也继承了 ICommand。也就是说继承 ITool 接口类一般都会继承 ICommand，以此满足使用 OnCreat 传参数的需求。这也就解释了以上代码中为什么我们使用 ICommand 定义自定义的工具，最终使用查询接口的方式将由 ICommand 转换到 ITool。详细代码可联系作者索取。

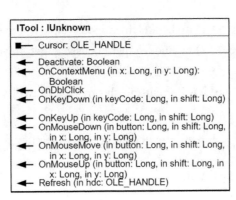

图 2-5-2　ITool 接口

4. ICommandSubtype

有些时候一些命令或工具等是可以分类的，例如，设置所有图层可见与设置所有图层不可见两个方法的代码区别很小，可以合并为同一类型命令，用几个变量进行区分。为此我们可以使用一个类封装两个命令，使用 Subtype 来控制命令的种类。以图层可见性为例，此时我们需要在继承 BaseCommand（或 ICommand）的基础上再继承 ICommandSubtype 接口（图 2-5-3），以获得 Subtype 属性（接口支持多继承，类只支持单继承）：

public sealed class LayerVisibility : BaseCommand, ICommandSubType

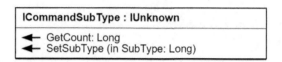

图 2-5-3　ICommandSubtype 接口

我们需要设置 GetCount 和 SetSubType 方法，此时需要注意的是，由于继承的是接口，设置这两个方法时不需要 override 关键字。可参考如下代码：

```
//设置 GetCount 方法获得共有几类命令
public int GetCount()
{
    return 2;
}
//设置 SetSubType 方法获得具体命令类型
public void SetSubType(int SubType)
{
    m_subType = SubType;
}
```

此时的 OnClick 事件也需要响应两类命令，参考代码如下：

```
for (int i=0; i <= m_hookHelper.FocusMap.LayerCount - 1; i++)
{
    if (m_subType == 1) m_hookHelper.FocusMap.get_Layer(i).Visible = true;
    if (m_subType == 2) m_hookHelper.FocusMap.get_Layer(i).Visible = false;
}
m_hookHelper.ActiveView.PartialRefresh(esriViewDrawPhase.esriViewGeography,null, null);
```

Caption 属性是命令的名称，相应地也应与命令类型同步变换，参考代码如下：

```
public override string Caption
{
    get
```

```
            {
                if (m_subType == 1)     return "Turn All Layers On";
                else      return "Turn All Layers Off";
            }
        }
```

另外，当所有图层处于可见状态时，Turn All Layers On 选项应为不可点击，Turn All Layers Off 选项应为可点击；当所有图层处于不可见的状态时，Turn All Layers On 选项应为可点击，Turn All Layers Off 选项应为不可点击。

参考代码如下：

```
            public override bool Enabled
            {
                get
                {
                    bool enabled = false; int i;
                    if (m_subType == 1)
                    {
                        for (i=0;i<=m_hookHelper.FocusMap.LayerCount - 1;i++)
                        {
                            if(m_hookHelper.ActiveView.FocusMap.get_Layer(i).Visible == false)
                            {
                                enabled = true;
                                break;
                            }
                        }
                    }
                    else
                    {
                        for (i=0;i<=m_hookHelper.FocusMap.LayerCount - 1;i++)
                        {
                            if (m_hookHelper.ActiveView.FocusMap.get_Layer(i).Visible == true)
                            {
                                enabled = true;
                                break;
                            }
                        }
                    }
                    return enabled;
```

         }
    }

自主练习：自行对照 OMD 查看 SystemUI 库中其他与命令、工具相关的接口，并尝试使用。

5. 命令封装与右键菜单

右键菜单的实现需要三个要素：①反馈点击要素，在之前控件的使用中介绍过，对于 TOCControl 控件可以使用 HitTest 方法；②当右击时，弹出菜单；③一定的命令封装，有些命令针对地图设置，有些命令针对图层设置，还有些命令针对图例设置。下面以 TOC 响应右键时的图层和地图的右键菜单为例设置右键菜单，已经封装的命令有 LayerVisibility、LayerSelectable、RemoveLayer、ZoomToLayer 和 ScaleThresholds，分别用于设置图层可见性、图层可选性、删除图层、缩放到图层、依据比例尺设置显示阈值。命令封装方法参照实验 2-4 中的内容。

1) 右键菜单定义

地图和图层的右键菜单弹出时有不同的响应，是因为分别进行了菜单设置。可以承载右键菜单的有 VS.Net 平台的 ContextMenu 和 ArcGIS Engine 平台的 ToolbarMenu，以下是使用 ToolbarMenu 分别进行地图和图层右键菜单设置的源码，使用 ToolbarMenu 的原因是其封装得更彻底。ToolbarMenu 使用 AddItem、AddSubMenu 等方法添加命令、工具等，最后需要设置钩子来传递操作对象。

相应参考代码如下：

```
//定义地图响应右键菜单
private IToolbarMenu m_menuMap;
//定义图层响应右键菜单
private IToolbarMenu m_menuLayer;
//Add custom commands to the map menu
m_menuMap = new ToolbarMenuClass();
m_menuMap.AddItem(new LayerVisibility(), 1, 0, false,
    esriCommandStyles.esriCommandStyleTextOnly);
m_menuMap.AddItem(new LayerVisibility(), 2, 1, false,
    esriCommandStyles.esriCommandStyleTextOnly);
//Add pre-defined menu to the map menu as a sub menu
m_menuMap.AddSubMenu("esriControls.ControlsFeatureSelectionMenu", 2, true);
//Add custom commands to the map menu
m_menuLayer = new ToolbarMenuClass();
m_menuLayer.AddItem(new RemoveLayer(), -1, 0, false,
    esriCommandStyles.esriCommandStyleTextOnly);
m_menuLayer.AddItem(new ScaleThresholds(), 1, 1, true,
    esriCommandStyles.esriCommandStyleTextOnly);
m_menuLayer.AddItem(new ScaleThresholds(), 2, 2, false,
    esriCommandStyles.esriCommandStyleTextOnly);
```

```csharp
m_menuLayer.AddItem(new ScaleThresholds(), 3, 3, false,
    esriCommandStyles.esriCommandStyleTextOnly);
m_menuLayer.AddItem(new LayerSelectable(), 1, 4, true,
    esriCommandStyles.esriCommandStyleTextOnly);
m_menuLayer.AddItem(new LayerSelectable(), 2, 5, false,
    esriCommandStyles.esriCommandStyleTextOnly);
m_menuLayer.AddItem(new ZoomToLayer(), -1, 6, true,
    esriCommandStyles.esriCommandStyleTextOnly);
//Set the hook of each menu
m_menuLayer.SetHook(m_mapControl);
m_menuMap.SetHook(m_mapControl);
```

2）右击响应，弹出菜单

右击响应依靠 HitTest 方法，HitTest 之后可以判断点击的是地图还是图层。依据点击的要素调用不同 ToolbarMenu 的 PopupMenu 方法，值得提醒的是，缩放到图层（ZoomToLayer）的命令是依靠 MaoControl 的 CustomProperty 属性获得 HitTest 反馈的具体图层对象的。

参考代码如下：

```csharp
//Ensure the item gets selected
if (item == esriTOCControlItem.esriTOCControlItemMap)
    m_tocControl.SelectItem(map, null);
else
    m_tocControl.SelectItem(layer, null);
//Set the layer into the CustomProperty (this is used by the custom layer commands)
m_mapControl.CustomProperty = layer;
//Popup the correct context menu
if (item == esriTOCControlItem.esriTOCControlItemMap) m_menuMap.PopupMenu
    (e.x, e.y, m_tocControl.hWnd);
if (item == esriTOCControlItem.esriTOCControlItemLayer) m_menuLayer.PopupMenu
    (e.x, e.y, m_tocControl.hWnd);
```

自主练习：对比 ContextMenu 实现右键菜单，实现其他功能，设计个性化、专业化程序。

6. MapControl 与 PageLayoutControl 官方同步

ArcGIS Engine 的官方示例程序中有一个 MapControl 和 PageLayoutControl 视图内容同步的例子：..\ArcGIS\DeveloperKit10.2\Samples\ArcObjectsNet\MapAndPageLayoutSynchApp\CSharp。其同步是靠 MapDocument 读取地图文档实现的，为此程序中的新建地图文档、打开地图文档、保存地图文档和另存为地图文档响应的代码都围绕着 MapControl 和 PageLayoutControl 视图内容同步来设计，并且封装了一个用于同步的 OpenNewMapDocument 的命令、一个 ControlsSynchronizer 类用于同步和 Maps 类用于添加删除地图操作。

自主练习：对比不同的同步方法，测试有哪些漏洞，回答应该怎么弥补？

实验 2-6　空间可视化

地图的基本语言，由形状不同、大小不一、色彩各异的图形和文字组成。空间可视化的过程就是利用这些符号、颜色、文字将地理空间数据最优表达的过程，它能够以直观的方式表达地理实体的空间位置、形状、质量、数量和地理实体间相互关系等特征。

（1）实验目的：了解 Display 库的符号、颜色体系，学会使用符号化控件。
（2）相关实验：实验 2-5　命令封装与右键菜单。
（3）实验数据：ArcGIS Engine 自带的示例数据或本教材系列实验数据。
（4）实验环境：Visual Studio 2012、ArcGIS Engine 10.2 和 C#语言。
（5）实验内容：颜色体系；符号；符号化控件使用；地图渲染。

1. 颜色体系

1）颜色

颜色（color）是符号和地图元素的基本属性，它组成了颜色元素（ColorElement）和颜色带（ColorRamp）。颜色有五种模型：①RGB 颜色模型，通过红色（red）、绿色（green）、蓝色（blue）三原色混合来显示；②CMYK 颜色模型，包括青（cyan）、洋红（magenta）、黄（yellow）和黑色（black，K），是减色模式，主要用于印刷中；③HSV 颜色模型，包括色调（hue）、饱和度（saturation）和颜色值或亮度（value）；④Gray 模型，使用 256 级的灰色来模拟颜色层次；⑤HLS 模型，包括色调（hue）、亮度（lightness）和饱和度（saturation）。

这五个颜色模型的组件类分别是：RgbColor、HsvColor、HlsColor、CmykColor 和 GrayColor，同时都有一个在类名前加 I 的接口。

以 HsvColor 为例设置颜色，参见以下代码，其中 Transparency 透明度为 0 时显示为透明，为 255 时显示为不透明。

```
IHsvColor phsv = new HsvColorClass();
phsv.Hue = 290;
phsv.Saturation = 50;
phsv.Value = 70;
phsv.Transparency = 255;
```

2）颜色对话框

ArcObjects 中提供了三种颜色对话框：颜色板（Colorpalatte）对象、颜色选择器（ColorSelector）对象和颜色浏览器（ColorBrowser）对象。这三个颜色对话框需要 Framework 库的支持，Framework 库本身就只有绑定 Desktop 权限产品时才可使用，因此这三个颜色对话框使用时需要绑定 Desktop 产品和许可。

（1）颜色板。颜色板（ColorPalette）对象一共排列了 120 种颜色供用户使用。ColorPalette 类实现了两个接口：IColorPalette 和 ICustomColorPalette。其中 IColorPalette 接口定义了 Color 属性和 TrackPopupMenu 方法，用于从对话框中获得颜色对象。

下面是使用调色板对象取出一个颜色并赋值给点的例子：

```
//设置初始默认颜色
IColor prgb = new RgbColorClass();
```

```
prgb.RGB = 255;
IColorPalette pPalette = new ColorPaletteClass();
//设置颜色板默认颜色
//定义一个范围结构
tagRECT pRect = new tagRECT();
pRect.left = 10;
pRect.top = 10;
pPalette.TrackPopupMenu(ref pRect, prgb, false, 0);
//获得新的颜色
IColor pOutColor = pPalette.Color;
//将颜色板中的颜色赋值给点
ICommand Pcom = new DrawPoint(pOutColor);
Pcom.OnCreate(m_mapControl.Object);
m_mapControl.CurrentTool = Pcom as ITool;
```

（2）颜色选择器。颜色选择器（ColorSelector）对象提供了一种更精确选择颜色的方法。用户可以点击对话框上方右边的小箭头，在 RGB、CMYK、HSV 多种颜色模型中选择，通过拖拽颜色带或者直接输入具体颜色值的方法返回颜色对象。ColorSelector 使用 DoModal 模态对话框来打开颜色选择器获得颜色。

参考代码如下：

```
//设置初始默认颜色
IColor prgb = new RgbColorClass();
prgb.RGB = 255;
IColorSelector pSelector = new ColorSelectorClass();
//设置颜色选择器默认颜色
pSelector.Color = prgb;
IColor pOutColor;
// Display the dialog
if (pSelector.DoModal(0))
{
    pOutColor = pSelector.Color;
}
else pOutColor = pSelector.Color;
//将颜色选择器中的颜色赋值给点
ICommand Pcom = new DrawPoint(pOutColor);
Pcom.OnCreate(m_mapControl.Object);
m_mapControl.CurrentTool = Pcom as ITool;
```

（3）颜色浏览器。颜色浏览器（ColorBrowser）对象提供了多种颜色模型（RGB、CMYK、HSV、HLS 或 Gray）供用户选择颜色对象。颜色浏览器（ColorBrowser）对象获得颜色的方法与颜色选择器（ColorSelector）对象一样，可参考以上代码。

3）颜色带

制作地图专题图时，往往需要随机或有序的颜色带（ColorRamp）来支持渲染工作。ArcGIS Engine 定义了 ColorRamp 类来生成颜色带，该类实现了 IColorRamp 接口，用于定义一系列颜色带的公共方法，如 Size（产生多少种颜色）、Colors（颜色带 IEnumColor）。ColorRamp 类是抽象类，由 Color 类组成，同时又组成了 ColorRampElement 类，继承该 ColorRamp 类的组件类有四种。

（1）AlgorithmicColorRamp 通过起止颜色来确定多个在这两个颜色之间的色带。该类实现了两个接口：IColorRamp 和 IAlgorithmicColorRamp，两个接口之间是接口继承关系，后者包含了前者所有的方法和属性。

（2）RandomColorRamp 使用 HSV 颜色模型来确定颜色范围，通过该范围定义随机颜色渐变。

（3）MultiPartColorRamp 定义多部分颜色渐变，叠加产生颜色带。

（4）PresetColorRamp 定义预设颜色渐变，可手动设置 13 种颜色。

ColorRamp 类往往与符号化控件和专题图渲染一起使用。

自主练习：尝试不同颜色设置方法。

2. 符号

ArcObjects 主要使用 MarkerSymbol、LineSymbol 和 FillSymbol 三种符号来绘制地理要素或图形元素的几何形状、使用 TextSymbol 进行文字标注、使用 3DChartSymbol 进行饼图等三维对象显示。所有的符号类都实现了 ISymbol 和 IMapLevel 接口，ISymbol 接口定义了符号对象的基本属性和方法；IMapLevel 接口的 MapLevel 属性可以调整符号的显示顺序。

ArcObjects10.2 版本中实现了 ISymbol 接口，见表 2-6-1。

表 2-6-1 实现了 ISymbol 的类及其描述

类名	描述
ArrowMarkerSymbol	依据预定义箭头创建的箭头符号
BarChartSymbol	定义条形图符号
CartographicLineSymbol	用于绘制实线或虚线的线符号
CharacterMarker3DSymbol(esri3DAnalyst)	3D 字体标记符号
CharacterMarkerSymbol	基于字体中字符的标记符号
ColorRampSymbol(esriCarto)	Esri ColorRampSymbol 用于栅格渲染
ColorSymbol(esriCarto)	Esri ColorSymbol 用于栅格渲染
DotDensityFillSymbol	点密度数据驱动的填充符号，需要点密度渲染器
GradientFillSymbol	填充符号由颜色渐变组成
HashLineSymbol	用于绘制散列或斜线的线符号
LineFillSymbol	填充符号由任何支持的线符号组成
Marker3DSymbol(esri3DAnalyst)	3D 标记符号组件
MarkerFillSymbol	填充符号由任何支持的点标记符号组成
MarkerLineSymbol	由点标记组成的线符号
MoleFillSymbol(esriDefenseSolutions)	摩尔填充符号类
MoleLineSymbol(esriDefenseSolutions)	摩尔线符号类

续表

类名	描述
MoleMarkerSymbol(esriDefenseSolutions)	摩尔点标记符号类
MultiLayerFillSymbol	包含一个或多个图层的填充符号
MultiLayerLineSymbol	包含一个或多个图层的线符号
MultiLayerMarkerSymbol	包含一个或多个图层的标记符号
PictureFillSymbol	基于 BMP 或 EMF 图片的填充符号
PictureLineSymbol	由 BMP 或 EMF 图片组成的线符号
PictureMarkerSymbol	基于 BMP 或 EMF 图片的点标记符号
PieChartSymbol	定义饼图符号
RasterRGBSymbol(esriCarto)	Esri RasterRGBSymbol 用于栅格渲染
SimpleFillSymbol	由一组预定义的样式组成的填充符号
SimpleLine3DSymbol(esri3DAnalyst)	简单的 3D 线符号组件
SimpleLineSymbol	由一组预定义样式组成的线符号
SimpleMarker3DSymbol(esri3DAnalyst)	简单的 3D 标记符号组件
SimpleMarkerSymbol	由一组预定义样式组成的点标记符号
StackedChartSymbol	定义堆积图表符号
TextMarkerSymbol(esriTrackingAnalyst)	对点几何进行符号化的文本标记符号
TextSymbol	控制文本显示方式的符号
TextureFillSymbol (esri3DAnalyst)	纹理填充符号组件
TextureLineSymbol (esri3DAnalyst)	纹理线符号组件

1）点符号

点符号（MarkerSymbol）对象是用于修饰点对象的符号，可以用于构建线、面符号，它拥有 6 个子类，其中不同的子类可以产生不同类型的点符号。所有的 MarkerSymbol 类都实现了 IMarkerSymbol 接口，这个接口定义了点状符号的公共方法和属性，如角度、颜色、大小和 XY 偏移量等。以 SimpleMarkerSymbol 为例，其设置点符号形状采用的是 ISimpleMarkerSymbol 的 Style 属性。

esriSimpleMarkerStyle 作为枚举类型还给出了其他的点符号，见表 2-6-2。

表 2-6-2 esriSimpleMarkerStyle 的类型

枚举常量	值	描述
esriSMSCircle	0	The marker is a circle
esriSMSSquare	1	The marker is a square
esriSMSCross	2	The marker is a cross
esriSMSX	3	The marker is an X
esriSMSDiamond	4	The marker is a diamond

以下设置为菱形：

    ISimpleMarkerSymbol pmarker = new SimpleMarkerSymbolClass();
    pmarker.Size = 2;

pmarker.Style = esriSimpleMarkerStyle.esriSMSDiamond;

2）线符号

线符号（LineSymbol）对象是用于修饰线型几何对象的符号。以 SimpleLineSymbol 为例，其设置线符号形状采用的是 ISimpleLineSymbol 的 Style 属性。以下利用枚举类型 esriSimpleLineStyle 设置为 DashDotDot 的形式，其他形式可查阅帮助文档：

  ISimpleLineSymbol Plinesymbol = new SimpleLineSymbolClass();
  Plinesymbol.Width = 1;
  Plinesymbol.Style = esriSimpleLineStyle.esriSLSDashDotDot;

3）面符号

面符号（FillSymbol）是用来修饰如多边形等具有面积的几何形体的符号对象，其实现方式为 IFillSymbol，这个接口定义了两个属性 Color 和 OutLine，以满足所有类型的 FillSymbol 对象的公共属性设置。

（1）IFillSymbol.Color 可以设置填充符号的基本颜色，当然如果不设置这个属性，则填充对象会显示默认颜色，其中大部分的填充对象都是黑色，GradientFillSymbol 为蓝色，LineFillSymbol 是中度灰色。

（2）IFillSymbol.OutLine 属性可以设置填充符号的外边框，这个外边框是一个线对象，因此使用 ILineSymbol 对象修饰。在默认情况下它是一个 Solid 类型的简单线符号。

以 SimpleFillSymbol 为例，其设置填充符号形状采用的是 ISimpleFillSymbol 的 Style 属性。以下利用枚举类型 esriSimpleLineStyle 设置为 BackwardDiagonal 的形式，其他形式可查阅帮助文档，而其边线符号需要一个线符号来支撑：

  ISimpleFillSymbol psimplefill = new SimpleFillSymbolClass();
  psimplefill.Style = esriSimpleFillStyle.esriSFSBackwardDiagonal;
  ISimpleLineSymbol pline = new SimpleLineSymbolClass();
  pline.Style = esriSimpleLineStyle.esriSLSDashDotDot;
  psimplefill.Outline = pline;

4）文字符号

文字符号（TextSymbol）对象是用于修饰文字元素的，文字元素在要素标注等方面很重要。TextSymbol 符号实现了三个主要的接口来设置字符：ITextSymbol、ISimpleTextSymbol 和 IFormattedTextSymbol。

（1）ITextSymbol 接口的 Font 属性是产生一个 TextSymbol 符号的关键。可以使用 VS.Net 平台中的 IFontDisp 接口来设置字体的大小和粗体、倾斜等属性，然后赋值给 ITextSymbol 接口的 Font 属性。ITextSymbol 接口还可以定义 TextSymbol 对象的颜色、角度、水平排列方式、垂直排列方式和文本等内容。

（2）ISimpleTextSymbol 接口主要用来设置一些简单属性，如 XOffset 和 YOffset 可以用于设置字符的偏移量，它还定义了一个重要的属性 TextPath，这个属性要传入一个 ITextPath 对象。

（3）IFormattedTextSymbol 接口的 ShaldowColor 属性可以设置阴影颜色，ShapeXOffset 和 ShapeYOffset 属性可以设置字体在 X 方向和 Y 方向上的偏移值，CharacterSpacing 和 CharterWidth 可以设置文本符号中单个字符之间的空隙和字符的宽度等属性。

以下为生成文字符号的核心代码：

```
ITextSymbol ptextsymbol =new TextSymbolClass();
ptextsymbol.Size = 10;
stdole.IFontDisp font=new stdole.StdFontClass() as stdole.IFontDisp;
ont.Name = "仿宋";
font.Size = 20;
ptextsymbol.Font = font;
ptextsymbol.Color.RGB = 200;
ITextElement ptextelement = new TextElementClass();
ptextelement.Text = "我喜欢 GIS 开发！！！ ";
ptextelement.Symbol = ptextsymbol;
IElement pele;
pele = ptextelement as IElement;
IPoint pt=new PointClass();
pt=activeView.ScreenDisplay.DisplayTransformation.ToMapPoint(X,Y);
pele.Geometry = pt;
IGraphicsContainer pGC = pmap as IGraphicsContainer;
pGC.AddElement(pele,0);
activeView.PartialRefresh(esriViewDrawPhase.esriViewGraphics, null, null);
```

5）3DChartSymbol

3DChartSymbol 是一个抽象类，它拥有三个子类：BarChartSymbol、PieChartSymbol 和 StackedChartSymbol。它本质上是一种 Marker 符号。

（1）BarChartSymbol 是最常用的三维着色符号，它使用不同类型的柱子来代表一个要素类中不同的属性，而柱子的高度取决于属性值的大小。该对象 IBarChartSymbol 接口的 VerticalBars 属性用于确定使用的柱子（Bar）是水平排列还是垂直排列，Width 和 Spacing 属性可以调节柱的宽度和柱之间的空隙，Axes 属性可以设置每根 Bar 的轴线，轴线是一个 ILineSymbol 对象，当 ShowAxes 为 True 时，能够显示轴线。

（2）PieChartSymbol 使用一个饼图来显示要素类的不同属性，不同的属性按照它们的数值大小占有一个饼图中的不同比例的扇形区域。该对象 IPieChartSymbol 接口的 ClockWise 属性用于确定饼图中颜色的方向，当 ClockWise 为 True 时，饼图中的颜色块呈顺时针方向分布。当 UseOutline 属性为 True 时，饼图的外框可以设置外框线：外框线使用 IPieChartSymbol.Outline 设置，它是一个 ILineSymbol 对象。

（3）StackedChartSymbol 也是 ChartRenderer 对象进行着色时最常用的符号，它使用的柱（StackedBar）是堆垒而成的。该对象使用 IStackedChartSymbol 接口设置 StackedChartSymbol 的外观，Width 属性用于设置柱的宽度，Outline 和 UseOutline 用于设置符号的外框线。当 Fixed 属性为 False 时，ChartRenderer 对象的每个 StackedBar 的尺寸会依据每个要素的属性来计算；当它为 True 时，则 StackedBar 的长度是一样的。

3DChartSymbol 的使用往往需要配合 ColorRamp 和符号化控件一起使用。

自主练习：编程实现多种符号化显示。

3. 符号化控件使用

ArcGIS 样式文件中保存了用于空间数据符号化的符号和地图元素。ArcGIS 内置的所有 Style/ServerStyle 文件位于＜ArcGIS 安装目录＞\Styles 文件夹中。

Desktop 的安装目录：..\ArcGIS\Desktop10.2\Styles 下包含了内置的 Style/ServerStyle 文件。

Engine 的安装目录：..\ArcGIS\Engine10.2\Styles 下包含了内置的 ServerStyle 文件。

ArcMap 程序最常使用的符号和地图元素都保存在 ESRI.style 文件中。Styles 文件夹中也有其他的 style 文件，如 Weather.style、Petrolenum.style 等，这些样式是为了满足不同行业的需求而制作的，使用时它们都需要被引入到 ArcMap 中。此外，用户可以创建自己的样式库。ArcObjects 开发人员可以使用 StyleGallery（仅适用绑定 Desktop 产品和许可）或 ServerStyleGallery（适用于 Engine、Desktop 和 Server）对象从 Style 或 ServerStyle 文件中取出样式符号供系统使用。

StyleGallery/ServerStyleGallery 类实现了 IStyleGallery 接口和 IStyleGalleryStorage 接口，这个接口提供了在 Stylegallery 对象中获得一个 Style 文件引用的方法，它也提供了方法让程序员能够添加或删除 Style 文件。

尽管使用 StyleGallery 或 ServerStyleGallery 对象能够选择系统中已经存在的样式符号，但是这种选择并不方便，不具备可视化图形界面让用户选择。SymbologyControl 控件，通过对话框以直观的方式来进行符号、样式的选择。

SymbologyControl 控件用于显示 ServerStyle 文件和 Style 文件的内容及自定义符号化，并可以选择单个符号进行图层的着色或设置元素的符号。在程序设计阶段，可以通过 SymbologyControl 的属性页，将 ServerStyle 文件载入 SymbologyControl 控件；也可以通过编程，使用 LoadStyleFile 方法将 ServerStyle 文件载入 SymbologyControl 控件，使用 RemoveFile 方法从控件中移除 ServerStyle 文件；使用 LoadDesktopStyleFile 方法将 Style 文件载入 SymbologyControl 控件。

SymbologyControl 一次只能显示一个 SymbologyStyleClass 中的内容，使用 ISymbologyControl 接口的 StyleClass 属性可以获取或设置当前的 SymbologyStyleClass，ISymbologyControl 接口的 GetStyleClass 方法返回特定的 SymbologyStyleClass。ISymbologyStyleClass 接口定义的属性方法用于管理 SymbologyStyleClass 中的样式条目（StyleGalleryItem）。使用该接口中的 RemoveItem、SelectItem、PreviewItem 方法可以移除、选择、预览单个的样式条目，使用 AddItem 方法可以添加定制的符号。SymbologyStyleClass 的类别由 esriSymbologyStyleClass 常量定义，该常量的取值包含 esriStyleClassMarkerSymbols、esriStyleClassLineSymbols、esriStyleClassFillSymbols、esriStyleClassNorthArrows、esriStyleClassScaleBars 等 22 个。

首先需要设计一个窗体 GetSymbol 用于承载符号化控件并添加相应的按钮、图片框、ComboBox 等进行交互响应。当该窗体弹出（Show 或者 ShowDialog）时，需要传递进来符号类型以设置符号化控件显示的默认符号。

以鼠标左键点击符号为例，核心代码如下：

```
private ISymbol GetSymbolByControl(ISymbol symbolType)
{
    ISymbol symbol = null;
    IStyleGalleryItem styleGalleryItem = null;
```

```csharp
            esriSymbologyStyleClass styleClass = 
esriSymbologyStyleClass.esriStyleClassMarkerSymbols;
            if (symbolType is IMarkerSymbol)
            {
                styleClass = esriSymbologyStyleClass.esriStyleClassMarkerSymbols;
            }
            if (symbolType is ILineSymbol)
            {
                styleClass = esriSymbologyStyleClass.esriStyleClassLineSymbols;
            }
            if (symbolType is IFillSymbol)
            {
                styleClass = esriSymbologyStyleClass.esriStyleClassFillSymbols;
            }
            GetSymbol symbolForm = new GetSymbol(styleClass);
            symbolForm.ShowDialog();
            styleGalleryItem = symbolForm.m_StyleGalleryItem;
            if (styleGalleryItem == null) return null;
            symbol = styleGalleryItem.Item as ISymbol;
            symbolForm.Dispose();
            this.Activate();
            return symbol;
        }
```

当选中的符号类型被传递到 GetSymbol 窗体后，会调用 GetSymbol 窗体的 Load 函数加载 Style/ServerStyle 的文件，以 ESRI.ServerStyle 的文件为例使用符号化控件进行加载，并从文件夹中读取所有的 Style/ServerStyle 的文件放置于 ComboBox 中，符号化控件从传递进来的符号类型获得默认显示符号类型，一般设置同类符号的第 1 个符号为默认符号。核心代码如下：

```csharp
            string sinstall = ESRI.ArcGIS.RuntimeManager.ActiveRuntime.Path;
            string DefaultStyleFile = System.IO.Path.Combine(sinstall, "Styles\\ESRI.ServerStyle");
            if (System.IO.File.Exists(DefaultStyleFile))
            {
                axSymbologyControl1.LoadStyleFile(DefaultStyleFile);
                axSymbologyControl1.StyleClass = StyleClass;
                axSymbologyControl1.GetStyleClass(axSymbologyControl1.StyleClass).SelectItem(0);
                comboBox1.Text = DefaultStyleFile;
            }
            comboBox1.Items.Clear();
```

```
StylePath = sinstall + "\\Styles";
string[] serverstyleFiles = System.IO.Directory.GetFiles(StylePath, "*.serverstyle",
System.IO.SearchOption.AllDirectories);
string[] styleFiles = System.IO.Directory.GetFiles(StylePath, "*.style",
System.IO.SearchOption.AllDirectories);
foreach (string file in serverstyleFiles)
{
    comboBox1.Items.Add(file);
}
foreach (string file in styleFiles)
{
    comboBox1.Items.Add(file);
}
```

实际上确定和取消按钮的主要作用就是隐藏该 GetSymbol 窗体，其中取消按钮的功能还有"设置类成员变量 m_StyleGalleryItem 为空"。真正为 m_StyleGalleryItem 赋值的是符号化控件的 OnItemSelected 方法，当某一符号被选择时触发 OnItemSelected 方法获得 styleGalleryItem，将该 styleGallery Item 赋值给类成员 m_StyleGalleryItem。核心代码参照如下：

```
m_StyleGalleryItem = e.styleGalleryItem as IStyleGalleryItem;
ISymbologyStyleClass symbologyStyleClass =
axSymbologyControl1.GetStyleClass(axSymbologyControl1.StyleClass);
/Preview an image of the symbol
stdole.IPictureDisp picture = symbologyStyleClass.PreviewItem(m_StyleGalleryItem,
pictureBox1.Width, pictureBox1.Height);
System.Drawing.Image image = System.Drawing.Image.FromHbitmap(new
System.IntPtr(picture.Handle));
pictureBox1.Image = image;
```

自主练习：编程实现 GetSymbol 窗体，加载不同的符号库文件，调用不同的符号类型。

4. 地图渲染

地图渲染（Render）位于 Carto 库中，有适用于矢量图层的 FeatureRenderer，适用于栅格图层的 RasterRenderer，适用于网络图层的 NetworedRenderer，以及适用于 Tin 图层、Las 图层和地形图层的 TinRemderer。这四个类都是抽象类，由其子类负责不同类型的着色运算。其渲染的基本思路是设置符号种类，然后依据渲染规则查找设定好的 ColorRamp 进行着色。

下面以唯一值符号化为例，首先需要设计一个窗体，该窗体用来辅助我们寻找渲染的图层和渲染的字段。选择好图层和字段之后，调用渲染函数，设置 IUniqueValueRenderer 接口对象，并将设置好的 IUniqueValueRenderer 对象赋值给 IGeoFeatureLayer 的 Renderer 属性，然后进行刷新。示例的核心代码如下：

```
private void Renderer()
{
    IGeoFeatureLayer pGeoFeatureL = (IGeoFeatureLayer)layer2Symbolize;
```

```csharp
IFeatureClass featureClass = pGeoFeatureL.FeatureClass;
//找出 rendererField 在字段中的编号
int lfieldNumber = featureClass.FindField(strRendererField);
if (lfieldNumber == -1)
{
    MessageBox.Show("Can't find field called " + strRendererField);
    return;
}
IUniqueValueRenderer pUniqueValueR = CreateRenderer(featureClass);
if (pUniqueValueR == null) return;
pGeoFeatureL.Renderer = (IFeatureRenderer)pUniqueValueR;
m_activeView.PartialRefresh(esriViewDrawPhase.esriViewGeography, null,
m_activeView.Extent);
}
```

设置 IUniqueValueRenderer 对象的方法为 CreateRenderer，该方法一方面判断选择图层的符号，另一方面调用生成 ColorRamp 方法获得 ColorRamp，利用循环按照字段个数将不同的颜色赋值给不同字段名的对象。代码为：

```csharp
private IUniqueValueRenderer CreateRenderer(IFeatureClass featureClass)
{
    int uniqueValuesCount = GetUniqueValuesCount(featureClass,
    strRendererField);
    System.Collections.IEnumerator enumerator = GetUniqueValues(featureClass,
    strRendererField);
    if (uniqueValuesCount == 0) return null;
    IEnumColors pEnumRamp =
    GetEnumColorsByRandomColorRamp(uniqueValuesCount);
    pEnumRamp.Reset();
    IUniqueValueRenderer pUniqueValueR = new UniqueValueRendererClass();
    //只用一个字段进行单值着色
    pUniqueValueR.FieldCount = 1;
    //用于区分着色的字段
    pUniqueValueR.set_Field(0, strRendererField);
    IColor pColor = null;
    ISymbol symbol = null;
    enumerator.Reset();
    while (enumerator.MoveNext())
    {
        object codeValue = enumerator.Current;
        pColor = pEnumRamp.Next();
```

```csharp
        switch (featureClass.ShapeType)
        {
            case ESRI.ArcGIS.Geometry.esriGeometryType.esriGeometryPoint:
                ISimpleMarkerSymbol markerSymbol = new SimpleMarker
                SymbolClass() as ISimpleMarkerSymbol;
                markerSymbol.Color = pColor;
                symbol = markerSymbol as ISymbol;
                break;
            case ESRI.ArcGIS.Geometry.esriGeometryType.esriGeometry
            Polyline:
                ISimpleLineSymbol lineSymbol = new SimpleLineSymbol
                Class() as ISimpleLineSymbol;
                lineSymbol.Color = pColor;
                symbol = lineSymbol as ISymbol;
                break;
            case ESRI.ArcGIS.Geometry.esriGeometryType. esriGeometry
            Polygon:
                ISimpleFillSymbol fillSymbol = new SimpleFillSymbolClass() as
                ISimpleFillSymbol;
                fillSymbol.Color = pColor;
                symbol = fillSymbol as ISymbol;
                break;
            default:
                break;
        }
        //将每次得到的要素字段值和修饰它的符号放入着色对象中
        pUniqueValueR.AddValue(codeValue.ToString(), strRendererField, symbol);
    }
    return pUniqueValueR;
}
```

以上代码使用了 RandomColorRampClass 进行 ColorRamp 的创建，参考代码如下：

```csharp
private IEnumColors GetEnumColorsByRandomColorRamp(int colorSize)
{
    IRandomColorRamp pColorRamp = new RandomColorRampClass();
    pColorRamp.StartHue = 0;
    pColorRamp.EndHue = 360;
    pColorRamp.MinSaturation = 15;
    pColorRamp.MaxSaturation = 30;
    pColorRamp.MinValue = 99;
```

```
    pColorRamp.MaxValue = 100;
    pColorRamp.Size = colorSize;
    bool ok = true;
    pColorRamp.CreateRamp(out ok);
    IEnumColors pEnumRamp = pColorRamp.Colors;
    pEnumRamp.Reset();
    return pEnumRamp;
}
```
唯一值符号渲染结果参见图 2-6-1。

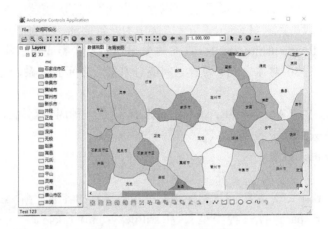

图 2-6-1 唯一值符号渲染图

自主练习：尝试实现不同图层的不同渲染方式。

实验 2-7 空间数据库

空间数据库是地理空间数据的存储、管理和分析计算的基础，ArcGIS Engine 统一采用 Geodatabase 模型进行数据的存储及管理。

（1）实验目的：学会 Geodatabase 相关程序设计。
（2）相关实验：实验 2-6 空间可视化。
（3）实验数据：ArcGIS Engine 自带的示例数据或本教材系列实验数据。
（4）实验环境：Visual Studio 2012、ArcGIS Engine 10.2 和 C#语言。
（5）实验内容：空间数据库相关操作功能实现。

1. Geodatabase 模型

Geodatabase 库定义了 Geodatabase 模型，主要对象如下。

（1）工作空间（Workspace）：代表一个 Geodatabase 或一个 ArcInfo Coverage 工作空间或一个文件夹（内有地理数据文件）。

（2）数据集（Dataset）：是任何数据的集合，可以是表（Table）、地理数据集（Geo Dataset）等。

（3）地理数据集（GeoDataset）：是一个包含了地理数据的数据集，可以分为要素数据集（FeatureDataset）和栅格数据集（RasterDataset）。

（4）要素数据集（FeatureDataset）：由要素类、几何网络和拓扑等组成。

（5）栅格数据集（RasterDataset）：在 Geodatabase 库中用专门的一页用来介绍栅格数据集，是与栅格数据相关的数据集。

（6）表（Table）：是数据库中的一个二维表，它由行（Row）组成，其列属性由字段集设置；Row 是表中的一条记录，一个表中的记录的字段集是相同的。

（7）对象类（ObjectClass）：是 Table 的扩展，它是一种具有面向对象特性的表，用于存储非空间数据；Object 代表了一个具有属性和行为的实体（entity）对象，而不是简单的 Row，具有 OID。

（8）要素类（FeatureClass）：是一种可以存储空间数据的对象类，它是对象类的扩展，其定义中包含几何字段；要素（Feature）是要素类中的一条记录，它是一个有几何字段的对象。

（9）关系类（RelationshipClass）：定义两个不同的要素类或对象类之间的关联关系；Relationship 代表关系类中对象之间、要素之间或对象要素之间的联系，它可以控制这些对象之间的行为。

（10）属性关系类（AttributedRelationshipClass）：是一种用于存储关系的表；Attributed Relationship 为属性关系类中的数据。

2. 工作空间及相关对象

1）工作空间

工作空间（Workspace）是空间数据集与非空间数据集的容器。由 esriWorkspaceType 枚举类型指定的 Workspace 有三种类型：①esriFileSystemWorkspace，是基于文件类型的 Shapefiles 和 ArcInfo 的工作空间；②esriLocalDatabaseWorkspace，是 Personal/File Geodatabase；③esriRemoteDatabaseWorkspace，是 ArcSDE Geodatabase。其他 Workspace 类型包括：①Raster Workspace，包含格网和影像数据的工作空间；②Tin Workspace，包含 TINs 数据的工作空间；③CAD Workspace，包含 CAD 数据的工作空间；④VPF Workspace，包含 VPF 数据的工作空间。

IWorkspace 接口定义了一个工作空间最普通的属性和方法。ConnectionProperties 返回工作空间的连接属性集对象；Datasets 属性可以按照数据集（Dataset）的类型而返回一个数据集枚举对象；Type 和 WorkspaceFactory 属性则可以分别返回工作空间的类型和工作空间工厂的种类。

IFeatureWorkspace 接口主要用于管理基于要素的数据集，如表（Table）、对象类（ObjectClass）、要素类（FeatureClass）、要素数据集（FeatureDataset）和关系类（Relationship Class）等。

2）工作空间工厂

要操作各种类型的空间数据，首先要获得空间数据所在的工作空间。工作空间是一个普通类，也就是说工作空间不能直接新建。往往需要使用具体的工作空间工厂（Workspace Factory）对象来创建或打开一个 Workspace。WorkspaceFactory 是 Geodatabase 的入口，但是它也是一个普通类，继承该类的有 ArcInfoWorkspaceFactory、ShapefileWorkspaceFactory、SdeWorkspaceFactory、AccessWorkspaceFactory、FileGDBWorkspaceFactory、RasterWorkspace Factory 等。不同类型的空间数据需要不同的工作空间工厂对象来打开对应的工作空间。

需要注意的是工作空间工厂的定义不在 Geodatabase 库中，例如，ArcInfoWorkspaceFactory 和 ShapefileWorkspaceFactory 的定义在 DataSourcesFile 库中；FileGDBWorkspaceFactory、AccessWorkspaceFactory 和 SdeWorkspaceFactory 的定义在 DataSourcesGDB 库中，Raster WorkspaceFactory 的定义在 DataSourcesRaster 库中。

IWorkspaceFactory 接口定义了所有工作空间工厂对象的一般属性和方法，用户可以通过它管理不同类型的工作空间，所有的工作空间对象都可以通过这个接口产生。IWorkspaceFactory.WorkspaceType 属性可以返回工作空间的类型。

（1）使用 IWorkspaceFactory.Create 方法可以用于新建一个工作空间名称对象。新建代码如下：

  public IWorkspaceName Create (string parentDirectory, string Name, IPropertySet ConnectionProperties, int hWnd);

（2）IWorkspaceFactory.Open 方法和 IWorkspaceFactory.OpenFromFile 方法可以用于打开一个已经存在的工作空间，可以调用如下代码：

  public IWorkspace Open (IPropertySet ConnectionProperties, int hWnd);

  public IWorkspace OpenFromFile (string fileName, int hWnd);

3）Propertyset 对象

Propertyset 对象是一个专门用于设置属性的对象，它是一种 name-value 对的集合。属性名必须是字符串，属性值可以是字符串、数值或日期，也可以是一个对象。属性值支持通过名字来查找属性的方法。IWorkspaceFactory.Open 方法要求用 Propertyset（属性集合）来设置打开一个 Workspace。如打开一个 SDE 数据库的时候，进行如下配置：

  IPropertySet pPropertyset = new PropertySetClass();

  pPropertyset.SetProperty("Server", "data") ;　　//服务器

  pPropertyset.SetProperty("Instance", "esrisde");　　//SDE 实例

  pPropertyset.SetProperty("user", "sde") ;　　//SDE 数据库的用户名

  pPropertyset.SetProperty("password", "sde");　　//SDE 数据库的密码

  pPropertyset.SetProperty("version", "sde.DEFAULT");　　//默认版本

4）名称对象

名称（Name）对象标识并且定义了 Geodatabase 对象（如数据集或者工作空间）或地图对象（如图层）。尽管名称对象只是它代表的对象的一个"代理"，但它支持程序员使用实例化的特定对象的 Open 方法。名称对象的子类众多，如 TableName、FeatureClassName、ObjectClassName 等。

下面是一个使用 Open 方法实例化的例子：

  IName pName = pFeatureClassName as IName;

  IFeatureClass pFeatureClass = pName.Open () as IFeatureClass;

5）AddData

SystemUI 库中实现了继承 ICommand 接口的 ControlsAddDataCommandClass 类，可以用于添加各种数据。参见以下代码：

  //添加到按钮中

  ICommand command = new ControlsAddDataCommandClass();

command.OnCreate(m_mapControl.Object);
command.OnClick();
//添加到右键菜单上
m_menuMap.AddItem(new ControlsAddDataCommandClass(), 0, 0, false, esriCommandStyles.esriCommandStyleIconAndText);

6）向工作空间添加 Shp 文件

添加 Shp 的基本思路是：

（1）定义一个 IWorkspaceFactory 接口，使用 ShapefileWorkspaceFactory 将该接口实例化。

（2）定义一个 IFeatureWorkspace 接口，使用实例化的 IWorkspaceFactory 接口的 OpenFromFile 将 Shp 文件打开，得到一个 FeatureWorkspace。

（3）定义一个 IFeatureClass 接口，使用 FeatureWorkspace 的 OpenFeatureClass 方法获得 FeatureClass。

（4）定义一个 IFeaturelayer，使用 FeaturelayerClass 实例化，将 FeatureClass 的信息赋值给 IFeaturelayer 得到 Featurelayer。

（5）使用 ActiveView 的 AddLayer 方法将 Featurelayer 添加到 MapControl 的 ActiveView 中，刷新 ActiveView。

具体参见以下代码：

```
// Create a new ShapefileWorkspaceFactory CoClass to create a new workspace
IWorkspaceFactory workspaceFactory = new ShapefileWorkspaceFactoryClass();
// System.IO.Path.GetDirectoryName(shapefileLocation) returns the directory part of the string Example: "C:\test\"
IFeatureWorkspace featureWorkspace = (IFeatureWorkspace)workspaceFactory.OpenFromFile(System.IO.Path.GetDirectoryName(shapefileLocation), 0); // Explicit Cast
// System.IO.Path.GetFileNameWithoutExtension(shapefileLocation) returns the base filename (without extension). Example: "cities"
IFeatureClass featureClass = featureWorkspace.OpenFeatureClass(System.IO.Path.GetFileNameWithoutExtension(shapefileLocation));
IFeatureLayer featureLayer = new FeatureLayerClass();
featureLayer.FeatureClass = featureClass;
featureLayer.Name = featureClass.AliasName;
featureLayer.Visible = true;
IActiveView activeView = axMapControl1.ActiveView;
activeView.FocusMap.AddLayer(featureLayer);
// Zoom the display to the full extent of all layers in the map
activeView.Extent = activeView.FullExtent;
activeView.PartialRefresh(esriViewDrawPhase.esriViewGeography, null, null);
```

7）新建个人数据库

新建个人数据库的基本思路是：

（1）定义一个 IWorkspaceFactory 接口，使用 AccessWorkspaceFactory 将其实例化。

（2）定义数据库的存储路径和名称。

（3）调用 WorkspaceFactory 的 Create 方法创建数据库，并将创建的数据库赋值给 IWorkspaceName。

（4）然后使用 IName 的 Open 函数打开 Workspace。

参考代码如下：

```
IWorkspace workspace = null;
IWorkspaceFactory workspaceFactory = new AccessWorkspaceFactoryClass();
string gdbName = "MyPersonalGeodatabase.mdb";
string newGDBDirectory = folderBrowserDialog1.SelectedPath;
string gdbFullPath = System.IO.Path.Combine(newGDBDirectory, gdbName);
IWorkspaceName workspaceName = workspaceFactory.Create(newGDBDirectory, gdbName, null, 0);
IName name = (ESRI.ArcGIS.esriSystem.IName)workspaceName;
workspace = (IWorkspace)name.Open();
MessageBox.Show("已在" + newGDBDirectory + "中创建了数据库 MyPersonalGeodatabase.mdb！");
```

8）新建文件型数据库

新建文件型数据库的基本思路与新建个人数据库的思路很像，差别在 WorkspaceFactory，即：

（1）定义一个 IWorkspaceFactory 接口，使用 FileGDBWorkspaceFactory 将其实例化。

（2）定义数据库的存储路径和名称。

（3）调用 WorkspaceFactory 的 Create 方法创建数据库，并将创建的数据库赋值给 IWorkspaceName。

（4）然后使用 IName 的 Open 函数打开 Workspace。

参考代码如下：

```
IWorkspace workspace = null;
IWorkspaceFactory workspaceFactory = new FileGDBWorkspaceFactoryClass();
string gdbName = "MyFileGeodatabase.gdb";
string newGDBDirectory = folderBrowserDialog1.SelectedPath;
string gdbFullPath = System.IO.Path.Combine(newGDBDirectory, gdbName);
IWorkspaceName workspaceName = workspaceFactory.Create(newGDBDirectory, gdbName, null, 0);
IName name = (ESRI.ArcGIS.esriSystem.IName)workspaceName;
workspace = (IWorkspace)name.Open();
MessageBox.Show("已在"+newGDBDirectory+"中创建了数据库 MyFileGeodatabase.gdb！");
```

自主练习：对照 OMD 查阅数据库的封装设计方式，思考其中的原因，并尝试其他数据库相关操作。

3. Dataset

所有放在工作空间中的对象都是一种数据集（Dataset）对象。Dataset 对象分为两大类：一种是 Table，可以简单看作一张二维表，它是由一条条记录组成的，是保存记录 Row 的容器；另一种是地理数据集（GeoDataset），如要素数据集、栅格数据集。

1）地理数据集

地理数据集是一个抽象类，它代表了拥有空间属性的数据集。GeoDataset 包括要素数据集 FeatureDataset、要素类 FeatureClass、TIN 和栅格数据集 RasterDataset 等。非 GeoDataset 的数据集包括 Table、对象类 ObjectClass 和关系类 RelationshipClass 等。

IGeoDataset 接口定义了 GeoDataset 对象的空间信息，包括空间参考和范围属性。通过 IGeoDataset.SpatialRefrence 可以获得一个 GeoDataset 对象的空间参考，IGeoDataset.Extent 则可以获得要素集的范围。

IGeoDatasetSchemaEdit 接口可以改变一个 GeoDataset 的空间参考，CanAlterSpatial-Reference 属性可以指示空间参考的可编辑性，AlterSpatialRefrence 方法可以重新设置与数据集关联的空间参考，该方法多用于给一个空间参考为 Unknown 的地理数据集设置空间参考。

2）FeatureDataset 对象

要素数据集对象是具有相同空间参考的要素类的容器。使用要素数据集的情况十分广泛，例如，几何网络、拓扑关系必须建立在一个要素数据集中。

在工作空间中对一个要素类进行编程的时候，需要注意这个要素类放在什么地方，是直接放在工作空间中，还是放在一个要素数据集中。当使用 IWorkspace 的 Datasets 属性来遍历一个工作空间内的数据集时，返回的只是直接放在工作空间的数据集，而保存在一个要素数据集中的要素类则不会被遍历。使用 IFeatureWorkspace 的 OpenFeatureClass 可以打开工作空间中的任何一个要素类，无论它是直接存放在工作空间还是存放在工作空间中的一个要素数据集中。

IFeatureDataset 接口继承 IDataset，其 CreateFeatureClass 方法可以用来在要素数据集中创建一个新的要素类。这个方法和 IFeatureWorkspace 的 CreateFeaureClass 的方法类似。新建要素类的空间参考是通过它的几何字段来设置的。

IFeatureClassContainer 接口用于管理要素数据集中的要素类。该接口的 ClassByName 和 Class(index)等属性都可以用来获取数据集中的特定的要素类。IFeatureClassContainer 的 ClassCount 和 Classes 属性分别可以获得要素数据集中的要素类的数目和得到一个要素类的枚举对象，ClassByID 属性可以让程序员通过对象类的 ID 值返回一个特定的对象类。

要素数据集中还可以存储关系类对象，可以通过 IRelationshipClassContainer 接口添加、新建和获得要素数据集中的关系类对象；该接口的 CreateRelationshipClass 方法可以在要素数据集中新建一个关系类，可调用的代码为：

  public IRelationshipClass CreateRelationshipClass (string relClassName, IObjectClass OriginClass, IObjectClass DestinationClass, string forwardLabel, string backwardLabel, esriRelCardinality Cardinality, esriRelNotification Notification, bool IsComposite, bool IsAttributed, IFields relAttrFields, string OriginPrimaryKey, string destPrimaryKey, string OriginForeignKey, string destForeignKey);

3）修改坐标系

修改图层坐标系的思路是：首先构建一个窗体交互获得图层，然后使用 IGeoDataset

SchemaEdit 的 CanAlterSpatialReference 属性查看该图层的空间参考是否可以编辑，如果可以编辑则将设定好的空间参考通过 IGeoDatasetSchemaEdit 的 AlterSpatialReference 方法进行修改。

参见以下代码：

```
IFeatureLayer pFL = GetFeatureLayer(layerName);
IFeatureClass pFeatureClass = pFL.FeatureClass;
IGeoDataset pGeoDataset = pFeatureClass as IGeoDataset;
IGeoDatasetSchemaEdit pGeoDatasetSE = pGeoDataset as IGeoDatasetSchemaEdit;
if (pGeoDatasetSE.CanAlterSpatialReference == true)
{
    ISpatialReferenceFactory2 pSpatRefFact = new SpatialReferenceEnvironmentClass();
    IGeographicCoordinateSystem pGeoSys = pSpatRefFact.CreateGeographicCoordinateSystem(4214);//esriSRGeoCSType .esriSRGeoCS_Beijing1954
    pGeoDatasetSE.AlterSpatialReference(pGeoSys);
    MessageBox.Show("已改变当前图层的空间参考！");
    m_activeView.Refresh();
}
else MessageBox.Show("当前图层的空间参考不能被改变！");
```

4）创建要素数据集

创建要素数据集的思路是：首先构建一个窗体交互打开一个数据库，获得其工作空间和预创建要素数据集的名称，然后使用 IFeatureWorkspace 的 CreateFeatureDataset 方法基于设定好的坐标系创建要素数据集。

参见以下代码：

```
IFeatureWorkspace pfeaworkspace = pworkspace as IFeatureWorkspace;
ISpatialReferenceFactory2 pSpatRefFact = new SpatialReferenceEnvironmentClass();
IGeographicCoordinateSystem pGeoSys = pSpatRefFact.CreateGeographicCoordinateSystem(4214);
pfeaworkspace.CreateFeatureDataset(featureDatasetName, pGeoSys);
```

5）创建要素类

创建要素类的思路是：首先构建一个窗体交互打开一个数据库获得其工作空间、数据集和预创建要素类的名称，然后使用 IFeatureDataset 的 CreateFeatureClass 方法基于设定好的字段、要素类种类等创建要素类。

核心代码如下：

```
featureClass = featureDataset.CreateFeatureClass(featureClassName, validatedFields, CLSID, null , ESRI.ArcGIS.Geodatabase.esriFeatureType.esriFTSimple, strShapeField, "");
```

6）修改数据源位置

修改数据源位置的思路是：首先构建一个窗体交互打开一个数据库获得其工作空间中的指定要素类作为数据源，然后选择需要更换的图层，将该数据源设定为该图层的数据源。主要使用 IFeatureWorkspace、IFeatureDataset、IFeatureLayer、IFeatureClass、ILayer、IDataset

和 IFeatureClassContainer 等接口。

自主练习：对照 OMD 查阅数据集的架构，并尝试其他数据集相关操作。

4. Table

从关系数据库的角度而言，Table 对象代表了 Geodatabase 里面的一张二维表，它是最简单的数据容器对象，存储的元素是 Row 对象，因此它也是一个数据集对象。Table 是 ObjectClass 和 FeatureClass 的父类。

所有的 Table 类（包括 Table、ObjectClass 和 FeatureClass）都实现了 IClass 接口，通过这个接口定义的属性和方法可以实现表对象的一般操作。IClass 的 Fields 可以获得一个表的字段结构，IClass.HasOID 属性则可以返回一个表是否有 OID 字段。而 IClass.OIDFieldName 属性返回 OID 字段的字段名；AddField 用于给表对象添加一个字段，AddIndex 用于给表添加一个索引；IClass.FindField 方法可以通过一个字段的字段名获得它在表的字段集中的索引号。

Table 主要实现了 ITable 接口，它继承自 IClass。这个接口定义的方法可以供用户查询、选择、插入、更新、删除表中的记录。

public IRow GetRow (int OID) 可以根据一条记录的 OID 值来获得 Row 对象本身。代码如下：

public ICursor GetRows (object oids, bool Recycling);

public ICursor Insert (bool useBuffering)用于得到一个具有插入记录功能的插入型游标，它和 CreateRowBuffer 产生的 RowBuffer 对象需要配合使用。

自主练习：对照 OMD 查阅表的架构，并尝试其他表相关操作。

实验 2-8　空间数据查询

通过空间数据查询快速实现空间数据选择、查询与统计是 GIS 数据操作的基本功能，也是特定数据子集进行再应用或空间分析的前提。

（1）实验目的：掌握空间数据查询的编程方法。

（2）相关实验：实验 2-7 空间数据库。

（3）实验数据：ArcGIS Engine 自带的示例数据或本教材系列实验数据。

（4）实验环境：Visual Studio 2012、ArcGIS Engine 10.2 和 C#语言。

（5）实验内容：图层属性查询；按照属性查询要素；按照位置查询要素；按照选定区域查询要素。

1. Cursor 与 FeatureCursor 对象

游标（Cursor）对象是一个指向数据的指针，本身不包含数据内容，是提供一个到 Row 对象或要素对象的链接。游标（Cursor）有三种类型：查询游标（Search 方法）、插入游标（Insert 方法）和更新游标（Update 方法）。Cursor 对象支持的接口是 ICursor。

三种类型游标的调用代码如下：

//Insert 方法返回一个插入型游标，它通常用于往表中插入一批记录。

ICursor　pCursor = ITable.Insert(　)

//Update 方法会返回一个更新型游标，它用于更新或者删除一批记录。

ICursor　pCursor = ITable.Update(　)

//使用 ITable.Search 方法对表进行查询后，可以得到一个查询型 Cursor 对象，它指向一个或多个 Row 对象。

ICursor pCursor= ITable.Search( )

FeatureCursor 是 Cursor 的一个子类，指向一个或多个要素。它实现了 IFeatureCursor 接口，这个接口定义的属性和方法与 ICursor 类似。

2. QueryFilter 与 SpatialFilter 对象

在关系数据库中，查询条件是使用 SQL 语句的 WHERE 子语句完成的。ArcObjects 中并不直接使用 SQL 语句，它通过多个不同对象的配合来实现数据选择的目的。QueryFilter 与 SpatialFilter 对象就承担了这个任务，QueryFilter 基于属性值或关系来选择数据，SpatialFilter 通过空间关系来选择数据。

1）QueryFilter

其示例代码为：

```
IQueryFilter pQueryFilter = new QueryFilterClass();
//设置过滤器对象的属性
pQueryFilter.WhereClause = "STATE_NAME = 'California'";
IFeatureSelection pFeatureSelection = pFeatureLayer as IFeatureSelection;
//如果在选择的过程中不过滤任何记录，而是返回所有的数据，则可以使用关键字"null"来替代 pQueryFilter 对象
pFeatureSelection.SelectFeatures(pQueryFilter,
esriSelectionResultEnum.esriSelectionResultNew, false);
//得到要素类中的要素个数
pFeatureClass.FeatureCount(null)
```

数据源不同时，WhereClause 的语法也不同，可以使用 Workspace 的 ISQLSyntax 接口来确定数据源使用的 SQL 语法，如表名、字段名使用的分隔符，引号使用的字符等。WhereClause 具体区别如下：

（1）字段名引用不同。如果要查询的数据源为 File Geodatabase、Shapefile、dBASE table、coverage 或 INFO 表，则字段名使用双引号引用，如"AREA"；如果要查询的数据源为 personal Geodatabase，则字段名使用方括号引用，如[AREA]；如果要查询的数据源为 ArcSDE Geodatabase 或 ArcIMS 影像服务或要素服务，则字段名直接使用，如 AREA；如果要查询的数据源为 Excel 文件（xls）中 worksheet 或文本文件（txt），一般情况下，字段名使用单引号引用，如'AREA'；但如果是从表窗口中打开 Select By Attributes 对话框，则字段名使用方括号引用，如[AREA]。

（2）字符串使用区别。字符串总是使用单引号引用，如："STATE_NAME" = 'California'；personal Geodatabase 中的字段值不区分大小写，而 ArcSDE、File Geodatabase 及 Shapefile 中的字段值大小写敏感。可以使用转换函数将字符串统一转换为大写或小写；基于文件的数据源，使用 UPPER 函数将字符串中的所有字符转换为大写，使用 LOWER 函数将字符串中的所有字符转换为小写，如 UPPER("LAST_NAME") = 'JONES'；其他数据源有类似的转换函数，如 personal Geodatabase 中的 UCASE、LCASE。使用 LIKE 运算符（代替"="）构建基于部分字符串的查询，如"STATE_NAME" LIKE 'Miss%'。

（3）通配符（代表一个或多个字符的特殊符号）使用区别。对于文件类型的数据源，'%'代表其所在位置的一个或多个或无字符；'_'代表其所在位置的任意一个字符，如"NAME" LIKE 'Cath%'、"OWNER_NAME" LIKE '_atherine smith'；对于 personal Geodatabase，'*'代表任意数目的字符，'?'代表任意单个字符。对于连接表，要使用合适的通配符；如果查询仅用到目标表的字段，则使用目标表的通配符；如果查询仅用到连接表的字段，则使用连接表的通配符；如果查询涉及两边的字段，则使用'%'和'_'通配符。

（4）其他。Geodatabases 中的字段、Shapefile/dBASE 表及 coverage/INFO 表中的 date 字段都支持 Null 值。File Geodatabase 不支持 Distinct 关键字，建议使用 IdataStatistics. UniqueValues 返回某个字段上的唯一值。可以使用=、<>、>、<、>=、<=运算符查询数值型数据，如"POPULATION96" >= 5000。日期的查询语法依赖于数据源类型，如 personal Geodatabase 中[DATE_OF_BIRTH] = #01-09-2019 18:30:00#，File Geodatabase 中"DATE_OF_BIRTH" = date '2019-01-09 18:30:00'。

使用 ISQLSyntax.GetSpecialCharacter 方法返回数据源使用的分隔符的前缀和后缀等信息，该方法的语法如下：

  public string GetSpecialCharacter (esriSQLSpecialCharacters sqlSC);

esriSQLSpecialCharacters 取值如下：

esriSQL_WildcardManyMatch，值为 1，多字符通配符；

esriSQL_WildcardSingleMatch，值为 2，单字符通配符；

esriSQL_DelimitedIdentifierPrefix，值为 3，分隔符前缀；

esriSQL_DelimitedIdentifierSuffix，值为 4，分隔符后缀；

esriSQL_EscapeKeyPrefix，值为 5，Escape Key 前缀；

esriSQL_EscapeKeySuffix，值为 6，Escape Key 后缀。

2）SpatialFilter

IspatialFilter 继承 IqueryFilter，如果要查询"在 5km 内的餐厅有哪些？"，就要使用 IspatialFilter。用户可以进行的空间过滤范围非常广泛，如寻找与某个选择区域相覆盖的要素、寻找某个要素附近的对象。

IspatialFilter 接口包含了空间属性和普通属性，其中有两个必须属性 Geometry 和 SpatialRel；GeometryField 属性用于设置几何字段名；SearchOrder 属性用于设置选择顺序，默认先空间后非空间排序，当 esriSearchOrderSpatial 时，空间优先；当 esriSearchOrderAttribute 时，属性优先。

参考以下示例：

  IspatialFilter pFilter = new SpatialFilterClass();
  //设置空间过滤器的三个必须属性
  pFilter.Geometry = pPolygon;
  pFilter.GeometryField = "SHAPE";
  pFilter.SpatialRel = esriSpatialRelEnum.esriSpatialRelIntersects;
  IfeatureCursor pFeatureCursor = pFeatureClass.Search(pFilter, false);

3. QueryDef 对象

QueryDef 对象代表了数据库中基于一个或多个表、要素类进行的属性查询。通过这个对象，用户可以在多个表间建立连接，并且保证在这个连接基础上的查询可以实现。QueryDef 不能在 Shapefile 和 Coverage 数据中使用，仅可在 Geodatabase 数据库中使用。QueryDef 中的表必须放在一个工作空间内。QueryDef 不是一个组件类，其对象是使用 IfeatureWorkspace 的 CreateQueryDef 方法来产生的。其查询结果由 Cursor 对象返回，不与它们的父表相关联，不能使用 store 方法。IfeatureWorkspace 的 OpenFeatureQuery 方法会产生一个基于 QueryDef 对象的要素类，可以加入到 Map 中。

4. TableSort 对象

TableSort 对象是表的排序对象，它可以将查找的结果按照某个字段进行排序，然后返回所有符合条件的要素游标。TableSort 的必设属性有 Table 和 Fields。

5. 要素选择集

IfeatureClass、IfeatureLayer 都提供了以下两种不同的查询手段。

（1）Search 方法：返回的是一个指向数据的查询型 Cursor 对象。

（2）Select 方法：构造一个选择集（SelectionSet），然后在选择集中再得到选择集，显示的效果比前者好，即得到的要素选择集在 Map 上会高亮显示。

可以使用 SelectionType 属性来设置选择集获得方式：如果是 esriSelectionTypeIDSet，那代表选择集使用的是一个 OID 集合；如果是 esriSelectionTypeSnapShot，表明选择集使用的是保存在内存中的实际的行对象；如果是 esriSelectionTypeHybird，当选择数量少的时候选择集使用在内存中的行对象，当数量多时则使用 OID 集合。

6. 查询图层属性表

查询图层属性表实际上与以上内容关系很小，其思路是：通过右击选择图层，弹出查询属性表右键菜单；点击后弹出一个窗体，该窗体用一个 DataGridView 来承载图层的属性，通过构造数据集设置 DataGridView 的 Datasource 属性。

构造数据集的代码参考如下：

```
private int ConstructDataSet(IfeatureLayer pFeatLyr)
{
    IlayerFields pFeatlyrFields = pFeatLyr as IlayerFields;
    IfeatureClass pFeatCls = pFeatLyr.FeatureClass;
    int rows = 0;
    if (m_layerDataSet.Tables[pFeatLyr.Name] == null)
    {
        DataTable pTable = new DataTable(pFeatLyr.Name);
        DataColumn pTableCol;
        for (int I = 0; I <= pFeatlyrFields.FieldCount – 1; i++)
        {
            pTableCol = new DataColumn(pFeatlyrFields.get_Field(i).AliasName);
            pTable.Columns.Add(pTableCol);
            pTableCol = null;
```

```
            }
            IfeatureCursor features = pFeatLyr.Search(null, false);
            Ifeature feature = features.NextFeature();
            while (feature != null)
            {
                DataRow pTableRow = pTable.NewRow();
                for (int I = 0; I <= pFeatlyrFields.FieldCount – 1; i++)
                {
                    //pTableRow[i] = feature.get_Value(i);
                    if (pFeatlyrFields.FindField(pFeatCls.ShapeFieldName) == i)
                    {
                        pTableRow[i] = pFeatCls.ShapeType;
                    }
                    else
                    {
                        pTableRow[i] = feature.get_Value(i);
                    }
                }
                pTable.Rows.Add(pTableRow);
                feature = features.NextFeature();
            }
            rows = pTable.Rows.Count;
            m_layerDataSet.Tables.Add(pTable);
            System.Runtime.InteropServices.Marshal.ReleaseComObject(features);
        }
        return rows;
    }
```

7. 按照属性查询

按照属性查询是根据属性条件，对某个要素图层查询满足条件的地理要素。

ArcGIS Engine 需要开发人员自己设计界面来构造属性查询条件，其实现思路如下：

（1）设计一个通过属性查询的窗体。

（2）设置一系列按钮，并设置按钮响应。

（3）在窗体的 Load 事件中加载当前地图对象中的要素图层到图层下拉列表框 ComboBox 中，默认选择是第一个图层。

（4）将 ComboBox 中默认图层的字段读取出来放置于属性字段的 ListBox 里面，当点击应用时执行查询。

查询的核心代码参考如下：

```
            IQueryFilter pQueryFilter = new QueryFilter() as IQueryFilter;
```

```
IFeatureLayer pFeatureLayer;
pQueryFilter.WhereClause = textBoxWhereClause.Text;
pFeatureLayer = GetLayerByName(comboBoxLayers.SelectedItem.ToString()) as
IFeatureLayer;
pFeatureSelection = pFeatureLayer as IFeatureSelection;
int iSelectedFeaturesCount = pFeatureSelection.SelectionSet.Count;
pFeatureSelection.SelectFeatures(pQueryFilter, selectmethod, false);//执行查询
//本次查询后，如果查询的结果数目为 0，则认为本次未查询到结果
if ( pFeatureSelection.SelectionSet.Count == 0)
{
    MessageBox.Show("没有符合本次查询条件的结果！");
    return;
}
//如果复选框被选中，则定位到选择结果
if (checkBoxZoomtoSelected.Checked == true)
{
    IEnumFeature pEnumFeature = m_mapcontrol.Map.FeatureSelection as IEnum
    Feature;
    IFeature pFeature = pEnumFeature.Next();
    IEnvelope pEnvelope = new Envelope() as IEnvelope;
while (pFeature != null)
{
    pEnvelope.Union(pFeature.Extent);
    pFeature = pEnumFeature.Next();
}
m_mapcontrol.ActiveView.Extent = pEnvelope;
m_mapcontrol.ActiveView.Refresh();//刷新
}
else m_mapcontrol.ActiveView.PartialRefresh(esriViewDrawPhase.esriViewGeo
Selection, null, null);
double i = m_mapcontrol.Map.SelectionCount;
i = Math.Round(i, 0);//小数点后指定为 0 位数字
```

8. 按照空间位置、空间关系查询

按照空间位置、空间关系查询是根据查询的几何形状和空间关系，对所选择的要素图层进行查询。

空间查询的一般操作步骤如下。

（1）选择要查询的要素图层。

（2）选择用于查询的几何形状类型：点、线、矩形、圆和多边形。

（3）根据选择的几何形状类型，在图上选择或绘制用于查询的几何形状。

（4）选择空间关系，执行查询。

通过编码形式执行空间查询的实现思路如下。

（1）构建一个窗体。

（2）设定选择类型，有四种：构造新的选择集、添加当前选择集、从当前选择集去除、从当前选择集中选择构造新的选择集。

（3）选择待查询的要素图层的名称，保存在 CheckedListBox 中。

（4）选择空间关系，保存在 ComboBox 中。

（5）设置查询图层的名称。

（6）设置缓冲区距离。

9. 按照设定区域查询

按照设定区域查询的思路是首先利用一个图形跟踪器选择一个范围；将选取的区域作为参数赋值给 ISpatialFilter 接口，通过 IFeatureLayer 的 Search 方法获得选择集。调用属性窗体，将选择集的属性赋值给属性窗体的 DataGridView。

核心代码参考如下，其查询示例效果如图 2-8-1 所示。

```
//获取画定范围
IGeometry pGeometry = this.m_mapControl.TrackPolygon();
//获取选择集
List<IFeature> pFeatureList = new List<IFeature>();
ISpatialFilter pSpatialFilter = new SpatialFilterClass();
IQueryFilter pQueryFilter = pSpatialFilter as IQueryFilter;
pSpatialFilter.Geometry = pGeometry;
//pSpatialFilter.SpatialRel =esriSpatialRelEnum.esriSpatialRelContains;
pSpatialFilter.SpatialRel = esriSpatialRelEnum.esriSpatialRelIntersects;
IFeatureCursor pFeatureCursor = pFeatureLayer.Search(pQueryFilter, false);
//ISelectionSet
IFeatureSelection pFeatureSelection = pFeatureLayer as IFeatureSelection;
pFeatureSelection.SelectFeatures(pQueryFilter, esriSelectionResultEnum.esriSelectionResultNew, false);
IFeature pFeature = pFeatureCursor.NextFeature();
while (pFeature != null)
{
    pFeatureList.Add(pFeature);
    pFeature = pFeatureCursor.NextFeature();
}
pFList = pFeatureList;
//设置信息显示窗体中 DataGridView 的属性，设置行数 pFeatureList.Count+1 包括字段名哪一行即表头
pAttributeForm.dataGridView1.RowCount = pFList.Count + 1;
//设置列数
```

pAttributeForm.dataGridView1.ColumnCount = pFList[0].Fields.FieldCount;
//遍历第一个要素的字段用于给列头赋值（字段的名称）
for (int m = 0; m < pFList[0].Fields.FieldCount; m++)
{
    pAttributeForm.dataGridView1.Columns[m].HeaderText = pFList[0].Fields.get_Field(m).AliasName;
}
for (int i = 0; i < pFList.Count; i++)
{
    IFeature pFeature = pFList[i];
    {
        for (int j = 0; j < pFeature.Fields.FieldCount; j++)
            //填充字段值
            pAttributeForm.dataGridView1[j, i].Value = pFeature.get_Value(j).ToString();
    }
}

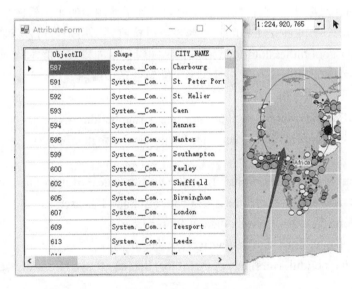

图 2-8-1　按照设定区域查询

自主练习：实现不同类型的空间查询，理解空间查询的含义。

实验 2-9　空间数据编辑

空间数据编辑是指产生新的矢量数据或对已有的矢量空间数据进行的再加工，包括矢量数据的生产、加工、维护、错误修正等，是空间数据维护与加工的基本环节。地理空间数据的几何形态往往需要空间数据编辑功能来支撑，空间数据编辑是 GIS 重要的功能之一。

（1）实验目的：学会空间编辑相关程序设计。

（2）相关实验：实验 2-8 空间数据查询。
（3）实验数据：ArcGIS Engine 自带的示例数据或本教材系列实验数据。
（4）实验环境：Visual Studio 2012、ArcGIS Engine 10.2 和 C#语言。
（5）实验内容：基本空间编辑功能的实现。

1. 编辑流程

ArcGIS 中的编辑操作由编辑会话辅助完成（使用插入游标或更新游标进行批导入或批更新要素除外）。一个编辑会话可以包含多个编辑操作（Edit Operation）和绘制操作（Sketch Operation）。编辑操作用于创建、修改或删除要素，而绘制操作用于修改编辑绘制（Edit Sketch）。编辑操作与绘制操作可以看作嵌套在编辑会话对应的长事务中的一系列短事务。通过把每个操作加入操作栈，实现编辑会话中的 Undo /Redo 功能。一旦编辑会话完成，所有操作都从操作栈删除。加入到操作栈的绘制操作是暂时的，绘制完成后，一系列的绘制操作将被一个编辑操作取代，编辑会话结束后，操作栈将被清空。使用 IEngineEditor（Controls 库）和 IWorkspaceEdit（Geodatabase 库）均可创建编辑会话。在不需要用户交互自动完成空间数据编辑功能的应用程序中，应使用 IWorkspaceEdit；在需要用户交互才能完成空间数据编辑的应用程序中，则使用 IEngineEditor。

使用 IEngineEditor.StartEditing 方法启动编辑会话，该方法有一个 IMap 类型和一个 IWorkspace 类型的参数。所有来自同一工作空间且处于地图中的可编辑图层均可在编辑会话中进行编辑；IEngineEditLayers.IsEditable 属性可以检测某图层是否可以编辑；在 SDE 工作空间中，IEngineEditor.EditSessionMode 可以指定编辑会话的模式（版本化或非版本化编辑），versioned 和 non-versioned 图层不能同时在一个编辑会话中编辑。

示例代码如下：

```csharp
private IEngineEditor m_engineEditor = new EngineEditorClass();
private void StartEditing (IMap map, IFeatureLayer featureLayer)
{
    if (m_engineEditor.EditState != esriEngineEditState.esriEngineStateNotEditing) return;
    IDataset dataset = featureLayer.FeatureClass as IDataset;
    IWorkspace workspace = dataset.Workspace;
    m_engineEditor.StartEditing(workspace, map);
    ((IEngineEditLayers) m_engineEditor).SetTargetLayer(featureLayer, -1);
}
```

IEngineEditor 接口 StartOperation 方法用于启动编辑操作；AbortOperation 方法用于取消编辑操作，不保存编辑；StopOperation 方法用于结束编辑操作，其需要 OperationName 参数传递指定操作栈。用于指定操作栈。

启动编辑操作代码参见：

```csharp
private void cmdEditOperation()
{
    m_engineEditor.StartOperation();
    try
    {
```

```
            //Perform feature edits here.
            if (someEditValidationChecksMethod == true)    m_engineEditor.Stop
            Operation("Test edit operation");
            else    m_engineEditor.AbortOperation();
        }
        catch (Exception ex)
        {
            m_engineEditor.AbortOperation();
            //Add code to handle exception.
        }
    }
```

通过执行 ControlsEditingSaveCommand 命令，可以保存编辑会话中的编辑变化，也可以调用 IEngineEditor 的 StopEditing 方法保存编辑变化；saveChanges 参数设置需设置为 true，若 saveChanges 参数为 false，则不保存编辑变化结束编辑会话。

```
    private void btnStopEditing_Click(object sender, EventArgs e)
    {
        if (m_engineEditor.HasEdits() == false)    m_engineEditor.StopEditing(false);
        else
        {
            if (MessageBox.Show("Save Edits?", "Save Prompt", MessageBox
            Buttons.YesNo) == DialogResult.Yes)    m_engineEditor.StopEditing(true);
            else    m_engineEditor.StopEditing(false)
        }
    }
```

IWorkspaceEdit 接口是 ArcObjects 实现空间数据编辑功能的另一接口，它不需要用户交互的编辑操作即可启动或停止一个编辑流程。使用 StartEditing 方法启动一个编辑流程，该方法需要一个 withUndoRedo 参数用来确定是否支持"undo/redo"功能。在启动编辑后，可以使用 StartEditOperation 方法开启编辑操作。如果在编辑过程中出现了异常，可以使用 AbortEditoperation 方法来取消编辑操作。在完成一个编辑后，用户可以使用 StopEditoperation 方法来确保编辑操作的完成。UndoEditoperation 方法可以用于编辑状态的回滚操作，如果发现编辑过程有误，通过执行这个方法可以恢复到最近变化前的状态。在整个编辑流程完成后，可以使用 StopEditing 方法来完成编辑。当执行完这个方法后，就意味着不能再进行"undo/redo"操作了。

使用 IWorkspaceEdit 可参考：

```
    private void StartEditing (IFeatureLayer featureLayer)
    {
        IFeatureClass featureClass = featureLayer.FeatureClass;
        IDataset dataset = featureClass as IDataset;
        IWorkspaceEdit workspaceEdit = dataset.Workspace as IWorkspaceEdit;
```

```
workspaceEdit.StartEditing(true);
workspaceEdit.StartEditOperation();
IFeature pFeature = featureClass.GetFeature(1);
pFeature.Delete();
workspaceEdit.StopEditOperation();
bool bHasEdits = true;
DialogResult iResponse = MessageBox.Show("Edit Operation", "Undo operation?", MessageBoxButtons.YesNo);
if (iResponse == DialogResult.Yes) workspaceEdit.UndoEditOperation();
workspaceEdit.HasEdits(ref bHasEdits);
if (bHasEdits)
{
    iResponse = MessageBox.Show("Edit Operation", "Save edits?", Message Box Buttons.YesNo);
    if (iResponse == DialogResult.Yes)
        { workspaceEdit.StopEditing(true); }
    else
        { workspaceEdit.StopEditing(false); }
}
}
```

2. 内置空间数据编辑命令

ArcGIS Engine 的 Controls 库中内置了空间数据编辑的命令，如 ControlsUndoCommand Class、ControlsRedoCommandClass、ControlsEditingStartCommandClass、Controls EditingStop CommandClass 和 ControlsEditingSaveCommandClass 等，可以添加到按钮命令中进行操作响应，也可以添加到 ToolbarControl 上。

参考示例代码如下：

```
//添加到按钮命令
ICommand command = new ControlsEditingStartCommandClass();
command.OnCreate(axMapControl1.Object);
command.OnClick();
//添加到 ToolbarControl
toolbarMenu.AddItem(new ControlsEditingStartCommand(), 0, 0, false, esriCommandStyles.esriCommandStyleTextOnly);
toolbarMenu.AddItem(new ControlsEditingStopCommand(), 0, 1, false, esriCommandStyles.esriCommandStyleTextOnly);
toolbarMenu.AddItem(new ControlsEditingSaveCommand(), 0, 2, false, esriCommandStyles.esriCommandStyleTextOnly);
toolbarMenu.AddItem(new ControlsEditingSnappingCommand(), 3, 0, true, esriCommandStyles.esriCommandStyleTextOnly);
```

ToolbarControl.AddItem(toolbarMenu, -1, -1, false, -1,
esriCommandStyles.esriCommandStyleTextOnly);

3. 设置编辑图层

在进行编辑时，需要明确对哪个图层进行编辑操作，所以在开始一个编辑对象之后，必须设置编辑哪一个图层。在 ArcGIS Engine 中设置编辑图层需使用 EngineEditor 类的 IEngineEditLayers 接口的相关方法和属性。EngineEditLayers 接口用于获取编辑会话过程中有关图层的信息，如判断一个图层是否可编辑、设置当前目标图层等。IEngineEditLayers 接口的主要方法和属性有：CurrentSubtype 属性，用于返回目标图层的子类型，若 IEngineEditLayers 的目标图层没有任何子类型，返回值为 0。TargetLayer 属性，用于返回 EngineEditor 的目标图层，目标图层就是通过命令或者编辑任务新建一个要素并写入的要素图层。IsEditable 方法，用于检查一个指定的图层是否可编辑，当开始一个编辑对象，用该方法自动访问地图中的每一个图层，只有可编辑的图层将被添加到 ControlsEditingTargetToolControl（编辑目标工具控件）中。SetTargetLayer 方法用于设置 EngineEditor 的目标图层。

核心代码参考如下：

//定义类成员变量

private IEngineEditLayers pEngineEditLayers = null;

pEngineEditLayers = pEngineEditor as IEngineEditLayers;

//设置编辑目标图层

pEngineEditLayers.SetTargetLayer(pCurrentLyr, 0);

4. 设置编辑任务

设置编辑任务主要是确定在编辑过程中执行什么样的操作，如添加要素、删除要素或修改要素。在 ArcGIS Engine 中设置编辑任务需使用 IEngineEditTask 接口的相关方法和属性。IEngineEditTask 接口负责向 EngineEditor 的 CurrentTask 属性传递当前所指定的编辑任务，这个任务也可以看作为了执行某个编辑操作（如创建要素等）而封装的一个流程。

实现设置编辑任务功能的核心代码如下：

//定义类成员变量

private IEngineEditTask pEngineEditTask = null;

//编辑任务实例化

pEngineEditTask = pEngineEditor as IEngineEditTask;

//通过名称获得任务

pEngineEditTask = pEngineEditor.GetTaskByUniqueName("ControlToolsEditing_CreateNewFeatureTask");

pEngineEditor.CurrentTask = pEngineEditTask;// 设置编辑任务

5. 编辑操作

编辑操作实现的功能主要包括选择要素、移动要素、添加要素、删除要素、撤销操作、恢复操作等。

选择要素是在当前地图窗口中进行要素选择的实现过程。

移动要素是指在编辑过程中，常常需要对某一个或者多个几何对象进行整体移动，以改变几何对象的位置。移动要素需要与鼠标配合使用，鼠标在控件上点击后触发 OnMouseDown

事件，获取移动的起始点；鼠标在控件上移动，触发 OnMouseMove 事件，实现移动要素到指定点；当释放鼠标左键时，完成移动要素事件。

添加要素需要 NewMultiPointFeedback、NewPolygonFeedback、NewLineFeedback 等对象的支持，分别用来添加多点、多边形、线等几何对象，三个类分别实现了 INewMultiPointFeedback、INewPolygonFeedback、INewLineFeedback 三个接口。添加要素也需要与鼠标配合使用，当鼠标在控件上点击时，触发 OnMouseDown 事件，获得第一个点坐标，添加第一个点，当图层类型为点层时，则直接在鼠标点击位置创建点要素对象；当鼠标在控件上移动时，创建的要素对象移动到鼠标移动点位置；当双击鼠标左键时，触发 OnDblClick 事件，获取所绘制的要素。

删除要素是用于删除已有不合理的要素，在删除过程中需要用到 IFeature 接口的 Delete 方法用于删除指定的要素。

撤销操作可以通过 IWorkspaceEdit2 接口的 UndoEditOperation 方法来恢复到错误发生前的状态。

恢复操作可以通过 IWorkspaceEdit2 对象的 RedoEditOperation 方法来恢复到误操作发生前的状态。

以选择要素为例，其核心代码为：

```
if (m_EngineEditor == null) return;
if (m_EngineEditor.EditState != esriEngineEditState.esriEngineStateEditing) return;
if (m_EngineEditLayers == null) return;
//获取目标图层
IFeatureLayer pFeatLyr = m_EngineEditLayers.TargetLayer;
IFeatureClass pFeatCls = pFeatLyr.FeatureClass;
//获取地图上的坐标点
IPoint pPt = m_activeView.ScreenDisplay.DisplayTransformation.ToMapPoint(X, Y);
IGeometry pGeometry = pPt as IGeometry;
double dLength = 0;
ITopologicalOperator pTopo = pGeometry as ITopologicalOperator;
//设置选择过滤条件
ISpatialFilter pSpatialFilter = new SpatialFilterClass();
//不同的图层类型设置不同的过滤条件
switch (pFeatCls.ShapeType)
{
    case esriGeometryType.esriGeometryPoint:
        //将像素距离转换为地图单位距离
        dLength = MapManager.ConvertPixelsToMapUnits(m_activeView, 8);
        pSpatialFilter.SpatialRel = esriSpatialRelEnum.esriSpatialRelContains;
        break;
    case esriGeometryType.esriGeometryPolygon:
        dLength = MapManager.ConvertPixelsToMapUnits(m_activeView, 4);
```

```
                pSpatialFilter.SpatialRel = esriSpatialRelEnum.esriSpatialRelIntersects;
                break;
            case esriGeometryType.esriGeometryPolyline:
                dLength = MapManager.ConvertPixelsToMapUnits(m_activeView, 4);
                pSpatialFilter.SpatialRel = esriSpatialRelEnum.esriSpatialRelCrosses;
                break;
        }
        //根据过滤条件进行缓冲
        IGeometry pBuffer = null;
        pBuffer = pTopo.Buffer(dLength);
        pGeometry = pBuffer.Envelope as IGeometry;
        pSpatialFilter.Geometry = pGeometry;
        pSpatialFilter.GeometryField = pFeatCls.ShapeFieldName;
        IQueryFilter pQueryFilter = pSpatialFilter as IQueryFilter;
        //根据过滤条件进行查询
        IFeatureCursor pFeatCursor = pFeatCls.Search(pQueryFilter, false);
        IFeature pFeature = pFeatCursor.NextFeature();
        while (pFeature != null)
        {
            //获取地图选择集
            m_Map.SelectFeature(pFeatLyr as ILayer, pFeature);
            pFeature = pFeatCursor.NextFeature();
        }
        m_activeView.PartialRefresh(esriViewDrawPhase.esriViewGeoSelection, null, null);
        System.Runtime.InteropServices.Marshal.ReleaseComObject(pFeatCursor);
```

6. 节点编辑

在实际的空间数据处理过程中，除添加、删除和移动几何对象这些针对整个几何对象的操作外，还经常需要对已经存在的几何对象进行形状的修改，这就需要对构成几何对象的节点进行编辑，包括添加节点、删除节点和移动节点等。

Polyline、Polygon 和 BezierCurve 这三类几何对象，都是由节点（vertex）构成其几何形状的。可分别用 LineMovePointFeedback、PolygonMovePointFeedback、BezierMovePointFeedback 这三个对象来移动这三类几何对象上的节点，它们分别实现了 ILineMovePointFeedback、IPolygonMovePointFeedback、IBezierMovePointFeedback 这三个接口。这三个接口移动节点的方法都是一样的，均由鼠标事件 MouseDown、MouseMove、MouseUp 来实现。如移动 Polyline、Polygon 上的节点时，当鼠标点击时，获取点击位置的节点，并根据图层类型自动判断所需用到的 Feedback 类型；当移动鼠标时，将选中节点移动到该点位置；当用户释放鼠标左键时重新绘制当前图形。添加节点类似，删除节点较为简单。

7. 属性编辑

在编辑过程中，往往需要对新建对象进行属性赋值或修改已有对象的属性信息。

Geodatabase 库的 IFeature 接口的 set_Value 方法可以用于给指定的字段赋值，在赋值完成之后需要用 Store 方法进行保存。属性编辑的实现步骤可参见"初始化窗体创建要素属性信息表"；将目标图层中被选中的要素在属性窗口中的图层名称列表中罗列；当鼠标点击图层树中的要素时，将所选要素的属性信息填入属性表；当属性表窗中的单元格的值发生变化时进行响应，需要时进行保存。

自主练习：实现自定义空间数据编辑功能，与内置空间数据编辑对比，思考如何通过修改自定义空间数据编辑的功能来弥补漏洞。

实验 2-10　空　间　分　析

丰富的空间分析功能是 GIS 软件的核心与特色。凭借各种空间分析工具，用户可以整合多种来源、不同格式的地理信息，进而执行大量、复杂的空间运算，挖掘数据隐含的信息。

（1）实验目的：掌握空间分析基本原理，学会设计空间分析程序。
（2）相关实验：实验 2-9 空间数据编辑。
（3）实验数据：ArcGIS Engine 自带的示例数据或本教材系列实验数据。
（4）实验环境：Visual Studio 2012、ArcGIS Engine 10.2 和 C#语言。
（5）实验内容：空间分析基础；地理处理工具（裁切、缓冲区分析、空间叠置分析）等。

1. 空间分析基础

1）空间关系运算

空间关系运算用于检测两个几何对象是否满足某种空间关系。Geometry 库中的 IRelationaloperator 接口定义了八种空间关系运算：Contains、Crosses、Disjoint、Equals、Overlaps、Relation、Touches 和 Within。实现 IRelationalOperator 接口的几何对象有 Point、Multipoint、Polyline、Polygon、MultiPatch、Envelope 和 GeometryBag。IRelationaloperator2 接口继承 IRelationaloperator 接口，另外又实现了三种空间关系运算：ContainsEx、IsNear 和 WithinEx。

以上空间关系运算时，需要指定基本几何对象和比较几何对象。基本几何对象是实例化 IRelationaloperator 接口的，比较几何对象是作为 IRelationaloperator 接口的输入参数。假设几何对象 A（pGeometryA）为基本几何对象（base geometry），几何对象 B（pGeometryB）为比较几何对象（comparision geometry），以 Contains 为例，则可以按如下方式使用该接口：

　　IRelationalOperator relOp = pGeometryA as IRelationalOperator;
　　bool yesOrNo = relOp.Contains(pGeometryB);

八种空间关系运算方法如下。

（1）Contains 方法：用于检测几何对象 A 是否包含几何对象 B，其包含计算情况，只有当几何对象 A 的维度大于或等于几何对象 B 的维度时才有包含情况,即点只可以包含点对象，多段线可以包含点和多段线，多边形可以包含点、多段线和多边形。

（2）Crosses 方法：用于检测几何对象 A 是否穿越了几何对象 B，有线穿越线、线穿越面、面穿越线三种情况。

（3）Disjoint 方法：用于检测几何对象 A 是否与几何对象 B 相分离，相分离的意思是两个几何对象的边界和内部都不相交。

（4）Equals 方法：可以检测两个几何对象是否相等，其要求是几何对象的形状、组成元素、属性均相同。也就是说只有点与点、线与线、多边形与多边形之间才可以有相等关系。不过处于相等关系的两个几何对象的 M 值和 Z 值可以不同。可以使用 IClone 的 IsEqual 方法来检测 M 值和 Z 值是否相同。

（5）Overlaps 方法：可以检测两个同种类型的几何对象是否有重叠。如果两类型几何对象 A、B 的交集与 A、B 的维数相同且与 A、B 都不相等，则说明两个几何对象有重叠。

（6）Relation 方法：用于检测几何对象 A 与 B 是否满足 relationDescription 指定的空间关系，当前版本只支持直线。

（7）Touches 方法：用于检测两个几何对象的相接关系。相接关系是指两个几何对象在它们的边界处相交，因此除了点与点外所有的类型的高级几何对象都可能拥有这种空间关系，如点与多段线相接于点、多段线与多段线和多边形相接于点、多段线与多边形相接于点、多边形与多边形相接于点或者多段线等。

（8）Within 方法：用于检测两个几何对象的相接关系和包含关系，基本与 Contains 的情况相反。

2）空间拓扑运算

使用空间拓扑运算可以基于一个或多个几何对象产生新的几何对象。ITopologicalOperator 接口定义了如下拓扑运算：Buffer（缓冲区运算）、Clip（用矩形裁切）、ClipDense（用矩形裁切，输出的线要求内插点加密）、ConvexHull（获得一个几何对象的最小凸多边形）、Cut（用线切割）、QueryClipped（用一个矩形去裁切一个几何对象）、QueryClippedDense（用一个矩形去裁切一个几何对象，输出的线要求内插点加密）、SymmetricDifference（两个几何对象的异或运算）、Intersect（两个几何对象的交集）、Difference（两个几何对象的差运算）、Union（两个几何对象的并集）、ConstructUnion（多个几何对象的并集）和 Simplify（简化几何对象）。

（1）Buffer 方法：可以给高级几何对象创建一个缓冲区多边形，无论是点、多段线还是多边形，它们的缓冲区都是一个具有面积的几何对象。

（2）Clip 方法：使用一个矩形去裁切一个几何对象，裁切结果为几何对象被矩形包围并去除了矩形之外部分。QueryClipped 方法也是用一个矩形去裁切几何对象，与 Clip 方法不同的是，QueryClipped 方法裁切的是其他几何对象，Clip 方法裁切的是自身几何对象。QueryClippedDense 方法与 QueryClipped 方法类似，ClipDense 方法与 Clip 方法类似，只是输出的边界线较密。

（3）ConvexHull 方法：可以产生一个几何对象的最小凸多边形，凸多边形是一个没有凹面且包含该几何图形的最小多边形。

（4）Cut 方法：用一条曲线去切割该几何对象，经过切割后把几何图形分为左、右两部分，左右两部分是相对曲线的方向而言的。

（5）Intersect 方法：可以返回两个同维度几何对象的交集，即两个对象的重合部分。

（6）Difference 方法：可以产生两个对象的差集。如果 A 是源对象，B 是参与运算的几何对象，则 C=A.Difference(B)，C 是 A 减去 A 与 B 的交集后剩下的部分。SymmetricDifference 方法是将 A 与 B 的并集减去 A 与 B 的交集的部分，即先将 A 与 B 进行 Union 运算，然后将 A 与 B 进行 InterSect 运算，将 Union 运算结果与 InterSect 运算结果做差。

（7）ConstructUnion 方法：可以将一个几何对象的枚举与同维度的单个几何对象合并，这种方法在大量几何对象需要合并的时候非常重要。Union 方法只可以合并两个同维度的单个几何对象，合并后的两个几何对象变成一个几何对象。

（8）Simplify 方法：可以让几何对象变得在拓扑上一致，拓扑规则如线不能自相交、多边形不能自重叠等。IsSimple 属性就是判断几何对象是否在拓扑上一致。

以下是使用 IRelationaloperator 接口的 Contains 方法判断两个多边形是否包含，如果包含则使用 ITopologicalOperator 接口的 Difference 方法进行差运算即把多边形挖空。其示例代码为：

```
IRelationalOperator pRelationalOperator = firstFeature.Shape as IRelationalOperator;
if (pRelationalOperator.Contains(secondFeature.Shape))
{
    ITopologicalOperator pTopologicalOperator = firstFeature.Shape as ITopologicalOperator;
    pGeometry = pTopologicalOperator.Difference(secondFeature.Shape);
    firstPolygonIsLarge = true;
}
```

ITopologicalOperator 接口还定义了三个属性：Boundary（几何对象的边界）、IsKnownSimple（是否已知几何对象为简单类型）、IsSimple（检测几何对象是否为简单类型）。Boundary 属性是几何对象的边界，多边形的边界线是多段线，多段线的边界是多点，点和多点的边界为空。实现 ITopologicalOperator 接口的几何对象有 Point、Multipoint、Polyline、Polygon、GeometryBag 和 MultiPatch。低等级的几何对象如 Segment、Path 或 Ring，可以组合成高级几何对象之后使用 ITopologicalOperator 接口。此外 ITopologicalOperator2～ITopologicalOperator6 定义了更多的拓扑运算相关操作，除了 ITopologicalOperator6 是独立的接口之外，其他的接口都是向前继承的接口，例如，ITopologicalOperator4 接口继承 ITopologicalOperator3 接口。

3）IProximityOperator 接口

IProximityOperator 接口用于邻近计算，主要有 QueryNearestPoint、ReturnNearestPoint 和 ReturnDistance 等方法。实现 IProximityOperator 接口的几何对象有 BezierCurve、CircularArc、EllipticArc、Envelope、Line、Multipoint、Point、Polygon 和 Polyline 对象。

（1）QueryNearestPoint 方法：是一种查询最近点的方式。用于查询该几何对象上离输入点最近的点；对于不同的线段延伸类型，这个最近点可以位于该几何对象的某个延伸位置上。

（2）ReturnNearestPoint 方法：用于查找并返回该几何对象上离输入点最近的点；对于不同的线段延伸类型，这个最近点可以位于该几何对象的某个延伸位置上。

（3）ReturnDistance 方法：返回两个几何对象之间的最短距离。

以下为返回两个几何对象最短距离的代码：

```
IPolyline plinemeasure;
plinemeasure = (IPolyline)m_mapControl.TrackLine();
ISpatialReferenceFactory spatialReferenceFactory;
spatialReferenceFactory = new SpatialReferenceEnvironment();
```

IProjectedCoordinateSystem pPCS;

pPCS = spatialReferenceFactory.CreateProjectedCoordinate System((int)esriSRProj CSType. esriSRProjCS_WGS1984N_AsiaAlbers);

plinemeasure.Project(pPCS);

m_mapControl.MapUnits = esriUnits.esriKilometers;

IGeometry input_geometry;

input_geometry = plinemeasure.FromPoint;

IProximityOperator proOperator = (IProximityOperator)input_geometry;

double check;

check = proOperator.ReturnDistance(plinemeasure.ToPoint);

MessageBox.Show("所测距离为：" + check.ToString("#######.##") + "米");

自主练习：实现多边形挖空和量测距离的方法，对照 Geometry 的 OMD 尝试利用 IRelationaloperator、ITopologicalOperator 和 IProximityOperator 接口对不同的几何对象进行空间分析运算的实验。

2. 地理处理工具

ArcGIS 提供的数百个地理处理工具组织在 19 个左右的工具箱内。在 ArcObjects 中，每个工具箱对应于一个类库，这些类库统一在 ESRI.ArcGIS.Geoprocessor 的命名空间中，见表 2-10-1。

Geoprocessing 库实现了地理处理框架和基础的地理处理工具，其最重要的接口是 GeoProcessor 类的 IGeoProcessor2，该接口中的 Execute 方法可以执行地理处理工具。另外 IGeoProcessorResult2 接口用于访问结果对象；IGPUtilities3 接口用于简化工作流的粗粒度的 ArcObjects 组件；IGpEnumList 接口提供 IGeoProcessor2 中的一些方法返回的 IGpEnumList；IGPServer2 接口用于地理处理服务；IGpValueTableObject 接口用于值表对象；IGPToolCommandHelper 接口用于从命令按钮执行工具；IDataElement 接口用于访问数据集的属性；IGPFieldMapping、IGPFieldMap 接口用于字段映射。IGPFunction2 和 IGPFunctionFactory 接口可以创建自定义的地理处理工具。另外，ArcObjects 中的 Geoprocessor 中也提供了 Geoprocessor 类，使用该类的对象也可执行地理处理工具。

表 2-10-1　工具箱与程序集对应关系

工具箱名称	命名空间
Analysis Tools	ESRI.ArcGIS.AnalysisTools
3D Analyst Tools	ESRI.ArcGIS.Analyst3DTools
Cartography Tools	ESRI.ArcGIS.CartographyTools
Conversion Tools	ESRI.ArcGIS.ConversionTools
Data Interoperability Tools	ESRI.ArcGIS.DataInteroperabilityTools
Data Management Tools	ESRI.ArcGIS.DataManagementTools
Editing Tools	ESRI.ArcGIS.EditingTools
Geocoding Tools	ESRI.ArcGIS.GeocodingTools
Geostatistical Analyst Tools	ESRI.ArcGIS.GeostatisticalAnalystTools

续表

工具箱名称	命名空间
Linear Referencing Tools	ESRI.ArcGIS.LinearReferencingAnalystTools
Multidimension Tools	ESRI.ArcGIS.MultidimensionTools
Network Analyst Tools	ESRI.ArcGIS.NetworkAnalystTools
Schematics Tools	ESRI.ArcGIS.SchematicsTools
Server Tools	ESRI.ArcGIS.ServerTools
Spatial Analyst Tools	ESRI.ArcGIS.SpatialAnalystTools
Spatial Statistics Tools	ESRI.ArcGIS.SpatialStatisticsTools
Tracking Analyst Tools	ESRI.ArcGIS.TrackingAnalystTools
Sample Tools	ESRI.ArcGIS.SampleTools
Parcel Fabric Tools	ESRI.ArcGIS.ParcelFabricTools

需要注意的是在 ESRI.ArcGIS.Geoprocessor 和 ESRI.ArcGIS.Geoprocessing 两个命名空间中都有 GeoProcessor 类。两者的区别是在编写程序时，Geoprocessor 中 GeoProcessor 类的关键字是 Geoprocessor，Geoprocessing 中 GeoProcessor 类的关键字是 GeoProcessorClass。

1）裁切（Clip）

裁切分析可以使用 Geoprocessor 中的 AnalysisTools 的 Clip 方法，语法如下：

Clip(object in_features, object clip_features, object out_feature_class)

其需要设置三个参数，分别是被裁切图层 in_features、裁切图层 clip_features 和输出图层 out_feature_class。

可以参照以下代码：

```
Geoprocessor gp = new Geoprocessor();
ESRI.ArcGIS.AnalysisTools.Clip clip = new ESRI.ArcGIS.AnalysisTools.Clip();
if (GetFeatureLayer(obj) != null)
{
    clip.in_features = GetFeatureLayer(obj);
    clip.out_feature_class = txtOutputPath.Text + "\\" + "clip_" + obj;
}
else
{
    return;
}
clip.clip_features = clipFeatureClass;
clip.cluster_tolerance = tolerance;
IGeoProcessorResult results = (IGeoProcessorResult)gp.Execute(clip, null);
```

2）缓冲区分析（Buffer）

缓冲区分析可以使用 Geoprocessor 中的 AnalysisTools 的 Buffer 方法，其语法如下：

Buffer_analysis (in_features, out_feature_class, buffer_distance_or_field, line_side, line_end_ type, dissolve_option, dissolve_field)

需要设置的参数是拟进行缓冲区分析图层 in_features、缓冲区距离和单位 buffer_distance_or_field、缓冲区方向 line_side、缓冲区边缘形状类型 line_end_type、是否聚合 dissolve_option、聚合字段 dissolve_field 和输出图层 out_feature_class。

实现的主要代码为：

```
Geoprocessor gp = new Geoprocessor();
gp.OverwriteOutput = true;
gp.AddOutputsToMap = true;
ESRI.ArcGIS.AnalysisTools.Buffer buffer = new Buffer();
IFeatureLayer bufferLayer = GetFeatureLayer(strBufferLayer);
buffer.in_features = bufferLayer;
bufferedFeatureClassName = strBufferLayer + "_" + "Buffer";
string outputFullPath = System.IO.Path.Combine(strOutputPath, bufferedFeature Class Name);
buffer.out_feature_class = outputFullPath;
buffer.buffer_distance_or_field = bufferDistanceField;
buffer.line_side = strSideType;
buffer.line_end_type = strEndType;
buffer.dissolve_option = strDissolveType;
buffer.dissolve_field = strDissolveFields;
IGeoProcessorResult results = (IGeoProcessorResult)gp.Execute(buffer, null);
return results;
```

3）空间叠置分析（Overlay）

根据选择的叠置分析类型（求交叠置、求和叠置、擦除叠置、同一性叠置、更新叠置、异或叠置），对选择的输入图层和叠置图层进行空间叠置分析。

以下是选择图层响应代码：

```
private void btnOverlay_Click(object sender, EventArgs e)
{
    if (strInputLayer == "" || strOverLayer == "") return;
    Geoprocessor gp = new Geoprocessor();
    gp.OverwriteOutput = true;
    gp.AddOutputsToMap = true;
    IGeoProcessorResult results = null;
    switch (strOveralyerType)
    {
        case "求交叠置":
            results = IntersectOverlay(gp);
            break;
        case "求和叠置":
```

```
                    results = UnionOverlay(gp);
                    break;
                case "擦除叠置":
                    results = EraseOverlay(gp);
                    break;
                case "同一性叠置":
                    results = IdentityOverlay(gp);
                    break;
                case "更新叠置":
                    results = UpdateOverlay(gp);
                    break;
                case "异或叠置":
                    results = SymDiffOverlay(gp);
                    break;
                default:
                    break;
            }
        }
```

（1）求交叠置。交集操作（Intersect）用于计算两个要素的几何交集，两个要素的公共部分保留，输入要素类的属性值将被复制到输出要素类。输入要素必须是简单要素：点、多点、线或面，还可以设置它们的精度等级。输入要素不能是复杂要素，如注记要素、尺寸要素或网络要素。其语法为：

Intersect_analysis (in_features, out_feature_class, join_attributes, cluster_tolerance, output_type)

其中，join_attributes 表示确定哪些属性要传递到输出要素类。ALL 表示输入要素与相关要素的所有属性（包括 FID）都将传递到输出要素类，该选项是默认设置。NO_FID 表示除 FID 外，输入要素与相关要素的其余属性都将传递到输出要素类。ONLY_FID 表示输入要素的所有属性与相关要素的 FID 将传递到输出要素类。

output_type 表示确定输出要素的类型。如果输入具有不同几何类型，则输出要素类几何类型默认与具有最低维度几何的输入要素相同。输出类型可以是具有最低维度几何或较低维度几何的输入要素类型。例如，如果所有输入都是面，则输出可以是面、线或点。如果某个输入类型为线但不包含点，则输出可以是线或点。如果任何一个输入是点，则输出类型只能是点。

可以参照如下代码：

```
        private IGeoProcessorResult IntersectOverlay( Geoprocessor gp)
        {
            IGpValueTableObject vtobject = new GpValueTableObjectClass();
            vtobject.SetColumns(1);           object row = null;
            row = GetFeatureLayer(strInputLayer);       vtobject.AddRow(ref row);
```

```
        row = GetFeatureLayer(strOverLayer);        vtobject.AddRow(ref row);
        IVariantArray pVarArray = new VarArrayClass();
        pVarArray.Add(vtobject);
        string outputFullPath = …
        pVarArray.Add(outputFullPath);
        pVarArray.Add(strJoinAttributeType);
        pVarArray.Add(tolerance);
        pVarArray.Add(strOutputFeatureType);
        IGeoProcessorResult results = gp.Execute("intersect_analysis", pVarArray, null) as
IGeoProcessorResult;
        return results;
    }
```

（2）求和叠置。求和叠置（Union）用于计算输入要素的几何并集。所有要素都将被写入输出要素类，且具有来自与其重叠的输入要素的属性。其语法为：

　　Union_analysis (in_features, out_feature_class, join_attributes, cluster_tolerance, gaps)

　　输入图层和叠加图层都必须是多边形图层，还可以设置它们的精度等级。可以参照以上求交叠置代码进行更改。

（3）擦除叠置。擦除叠置（Erase）用于擦除输入要素中被 erase 要素所覆盖的要素。其语法为：

　　Erase_analysis (in_features, erase_features, out_feature_class, cluster_tolerance)

　　擦除要素可以为点、线或面，只要输入要素的要素类型等级与之相同或较低。面擦除要素可用于擦除输入要素中的面、线或点；线擦除要素可用于擦除输入要素中的线或点；点擦除要素仅用于擦除输入要素中的点。可以参照以上求交叠置代码进行更改。

（4）同一性叠置。同一性叠置（Identity）操作通过计算输入要素与 Identity 要素的几何交集，将覆盖在输入要素范围内的要素保留下来。对于输入要素，与 Identity 要素重叠的部分，将获得 Identity 要素的属性。其语法为：

　　Identity_analysis (in_features, identity_features, out_feature_class, join_attributes, cluster_tolerance, relationship)

　　输入要素可以是多边形、线或点，Identity 要素必须是多边形，或与输入要素的几何类型相同。可参照以下代码：

```
        private IGeoProcessorResult IdentityOverlay( Geoprocessor gp)
        {
            Identity identity = new Identity();
            identity.in_features = GetFeatureLayer(strInputLayer);
            identity.identity_features = GetFeatureLayer(strOverLayer);
            string outputFullPath = System.IO.Path.Combine(strOutputPath, strInputLayer + "_"
+ strOverLayer + "_" + "Identity.shp");
            identity.out_feature_class = outputFullPath;
            identity.join_attributes = strJoinAttributeType;
```

identity.cluster_tolerance = tolerance;
//identity.relationship =;
IGeoProcessorResult results = (IGeoProcessorResult) gp.Execute(identity, null);
return results;
}

（5）更新叠置。更新叠置（Update）先计算两个多边形要素的交集，然后用 Update 要素更新它所覆盖的输入要素。其语法为：

Update_analysis (in_features, update_features, out_feature_class, keep_borders, cluster_tolerance)

输入要素和更新要素均为多边形要素。输入要素类与更新要素类的字段名称必须保持一致。如果更新要素类缺少输入要素类中的一个（或多个）字段，则将从输出要素类中移除缺失字段的值。

（6）异或叠置。异或叠置（SymDiff）先计算输入要素与更新要素的交集，删除公共部分，保留其余部分。其语法为：

SymDiff_analysis (in_features, update_features, out_feature_class, join_attributes, cluster_tolerance)

输入和更新要素类或要素图层必须具有相同的几何类型。

自主练习：实现以上地理处理工具，对照 Geoprocessor 和 Geoprocessing 两个命名空间里的其他对象实现其他空间分析运算。

实验 2-11　地图整饰输出

将数据可视化成果以地图的方式呈现出来是 GIS 基本的功能之一，编程开发整饰地图和输出是 GIS 应用开发不可或缺的环节。

（1）实验目的：掌握地图整饰输出相关编程能力。
（2）相关实验：实验 2-10 空间分析。
（3）实验数据：ArcGIS Engine 自带的示例数据或本教材系列实验数据。
（4）实验环境：Visual Studio 2012、ArcGIS Engine 10.2 和 C#语言。
（5）实验内容：地图布局整饰；Element 对象；MapGrid 对象；MapSurround 对象；地图输出。

1. 地图布局整饰

Carto 库定义了地图布局整饰相关的元素，如 PageLayout、Page、SnapGrid、SnapGuides 和 Rulersettings。

1）PageLayout

PageLayout 对象对应于 ArcMap 的布局视图，主要用于地图的设计、打印与输出，在其上放置和排列地理数据和地图元素，如数据框、地图标题、地图图例和地图比例尺等。PageLayout 对象通过 MapFrame 对象来管理文档中的地图对象，所有地图对象必须包含在 MapFrame 元素中，由 PageLayout 直接管理。PageLayout 会自动产生一些对象来修饰地图（SnapGuides、SnapGrid、RulerSettings 和 Page），以便更好地显示地图、打印和输出，其中打印机对应的是操作系统的打印机。

PageLayout 类主要实现了 IPageLayout3 接口，用于管理页面布局相关的页面、打印机、标尺、捕捉格网、水平捕捉线和垂直捕捉线对象，同时还提供了页面缩放、地图框遍历等方法。Page 属性用于管理 PageLayout 中的页面（Page）对象。RulerSettings、SnapGrid、HorizontalSnapGuides 和 VerticalSnapGuides 分别用于管理 PageLayout 上的标尺、捕捉格网、水平捕捉线和垂直捕捉线，这些对象有利于精确放置各种地图元素。ZoomToWhole 方法可以让 PageLayout 以最大尺寸显示；ZoomToPercent 可以按照输入的比例显示；ZoomToWidth 可以让视图显示的范围匹配控件对象的宽度。

参考示例代码如下：

```
private void ZoomPer()
{
    IPageLayout3    pPageLayout = axPageLayoutControl1.PageLayout;
    //页面缩小到 30%
    pPageLayout.ZoomToPercent(30);
}
```

PageLayout 对象通过 IGraphicsContainer 接口管理元素（Element）对象；使用 IGraphicsContainerSelect 接口管理被选择的元素；使用 IActiveView 和 IActiveViewEvents 接口管理页面布局的视图及相关事件。

2）Page

PageLayout 对象被创建后，会自动产生一个 Page 对象来管理布局视图中的页面，通过 IPageLayout3 的 Page 属性可以得到它的引用。

Page 类的 IPage 接口用于管理 Page 的颜色、尺寸、方向、版式单位、边框类型和打印区域等属性。IPageEvents 接口，用于管理页面的相关事件，如 PageColorChanged、PageMarginsChanged、PageSizeChanged 和 PageUnitsChanged 等，Page 对象会负责监听这些事件，并作出相应的反应。例如，当 Page 的单位发生变化后，布局视图需要更新它的转换参数、SnapGrid 及 SnapGuides 等附属对象。IPage 的 Orientation 属性为 2 时是横向，为 1 时是纵向，然而设置整个 PageLayout 布局的横纵向切换并不简单，因为它涉及的不仅仅是 Page 还有 Frame，另外还与打印机有关。IPage 的 FormID 可以设置纸张的类型或者设置与打印机相同的纸张，其由枚举类型 esriPageFormID 定义，见表 2-11-1。

表 2-11-1　枚举类型 esriPageFormID 的定义

枚举常量	值	描述
esriPageFormLetter	0	Letter - 8.5in×11in
esriPageFormLegal	1	Legal - 8.5in×14in
esriPageFormTabloid	2	Tabloid - 11in×17in
esriPageFormC	3	C - 17in×22in
esriPageFormD	4	D - 22in×34in
esriPageFormE	5	E - 34in×44in
esriPageFormA5	6	A5 - 148mm×210mm
esriPageFormA4	7	A4 - 210mm×297mm
esriPageFormA3	8	A3 - 297mm×420mm

续表

枚举常量	值	描述
esriPageFormA2	9	A2 - 420mm×594mm
esriPageFormA1	10	A1 - 594mm×841mm
esriPageFormA0	11	A0 - 841mm×1189mm
esriPageFormCUSTOM	12	自定义页面尺寸
esriPageFormSameAsPrinter	13	与打印机一致

* in 即英寸，1in≈25.4mm。

3）SnapGrid

SnapGrid 是 PageLayout 上为了摆放元素而设置的辅助点，这些点有规则地呈网状排列，便于用户对齐元素。SnapGrid 类实现了 ISnapGrid 接口，用于设置 SnapGrid 的属性。其中 HorizontalSpacing 和 VerticalSpacing 属性用于设置网点之间的水平距离和垂直距离，IsVisible 属性用于确定这些网点是否处于可见状态，Draw 方法用于将一个 SnapGrid 对象绘制在 Page 上。

其示例代码如下：

```
public void SnapGrid(IActiveView activeView)
{
    IPageLayout3 pageLayout = activeView as IPageLayout3;
    ISnapGrid snapGrid = pageLayout.SnapGrid;
    snapGrid.HorizontalSpacing = 5;
    snapGrid.VerticalSpacing = 5;
    snapGrid.IsVisible = true;
    activeView.Refresh();
}
```

4）SnapGuides

SnapGuides 是为了更好地放置地图元素而在 PageLayout 出现的辅助线。这个对象有两种类型：一种是水平辅助线，可通过 IPageLayout3 的 HorizontalSnapGuides 属性获得；另一种是垂直辅助线，可通过 IPageLayout3 的 VerticalSnapGuides 属性获得。每个 SnapGuides 都管理着一个 Guide 集合，即这种辅助线可以同时存在多条。SnapGuides 类实现了 ISnapGuides 接口，它定义了管理 SnapGuide 的属性和方法。其中 AreVisible 属性用于设定 SnapGuides 是否可见；GuideCount 属性可以返回一个 SnapGuides 对象中 Guide 的个数；Guide 属性可以按索引值得到某个具体的 Guide 对象；AddGuide 方法可以将一个 Guide 放在指定位置上；RemoveAllGuides 和 RemoveGuide 方法分别可以清除所有的 Guide 和按照索引值清除 Guide。

示例代码如下：

```
private void AddGuide (IActiveView activeView)
{
    IPageLayout3 pageLayout = activeView as IPageLayout3;
    ISnapGuides pSnapGuides = pageLayout.HorizontalSnapGuides;
    pSnapGuides.AddGuide(5);
```

```
                pSnapGuides.AddGuide(7);
                pSnapGuides.AddGuide(9);
                pSnapGuides.AddGuide(11);
                pSnapGuides = pageLayout.VerticalSnapGuides;
                pSnapGuides.AddGuide(2);
                pSnapGuides.AddGuide(4);
                pSnapGuides.AddGuide(6);
                pSnapGuides.AddGuide(8);
                pSnapGuides.AreVisible = true;
                activeView.Refresh();
          }
```
5）RulerSettings

标尺对象是为了辅助元素的放置而出现在 PageLayout 对象上方和左方的辅助尺，尽管 RulerSettings 是组件类，但它一般都是直接通过 IPageLayout3 的 RulerSettings 属性获得，获得与当前 PageLayout 相关的标尺。RulerSettings 实现了 IRulerSettings 接口，仅定义了一个属性 mallestDivision 用于设置最小的区分值（页面尺寸单位）。

2. Element 对象

Carto 库的 Element 是一个非常庞大复杂的对象集合，它主要分为两大部分：图形元素（Graphic Element）和框架元素（Frame Element）。

图形元素包括 GroupElement、MarkerElement、LineElement、TextElement、PictureElement 和 FillshapeElement 等对象，它们都是作为图形的形式而存在，在视图上是可见的。

框架元素包括 FrameElement、MapFrame、MapSurroundFrame、OleFrame 和 TableFrame 等对象，它们都是作为不可见的容器而存在的。IElement 是所有图形元素和框架元素类都实现的接口。这个接口可以确定元素的 Geometry 属性、查找元素和绘制元素。

图形元素在几何对象和空间参考章节已经有一定介绍，可自行查阅 OMD 进行学习，以下为框架元素介绍。

框架元素（Frame Element）是一种包含其他地图元素的容器。所有的图片元素都属于框架元素，除此之外，还有两个主要的框架元素：MapFrame（地图框架）和 MapSurroundFrame（地图附属物框架）。MapFrame（地图框架）对象是 Map 的容器，它用于管理 Map 对象；而 MapSurroundFrame 对象则用于管理 MapSurround 对象，MapSurround 就是为了修饰地图而使用的比例尺、比例文本、指北针、图例等对象。每个 MapSurroundFrame 都是与一个 MapFrame 相联系的。如果一个 MapFrame 被删除了，那么它其中所有的 MapSurroundFrame 对象也将被删除。框架元素类使用 IFrameElement 接口定义操作框架元素的属性和方法。

1）MapFrame 对象

MapFrame 由 PageLayout 控制，使用 IGraphicsContainer 接口的 FindFrame 方法可以查找到某个特定对象的框架对象。MapFrame 对象的 IMapFrame 接口的属性和方法用于控制其中的 Map 对象，该接口的只读属性 Map 可以获得这个地图框架内的地图对象，MapBounds 属性则可以返回地图对象的范围（Envelope），MapScale 属性可以确定地图显示的比例，CreateSurroundFrame 方法用于创建一个 MapSurroundFrame 对象。MapFrame 对象还实现了

IMapGrids 接口，它可以管理地图框架中的 MapGrid（地图格网）。

以下为改变 MapFrame 和 Page 以使 PageLayout 变成横向，代码为：

```
IMap pMap
IGraphicsContainer pGraphicsContainer;
IMapFrame pMapFrame;
pMap = axPageLayoutControl1.ActiveView.FocusMap;
pGraphicsContainer = (IGraphicsContainer)axPageLayoutControl1.PageLayout;
pMapFrame = (IMapFrame)pGraphicsContainer.FindFrame(pMap);
ISymbolBorder pSymborder = new SymbolBorderClass();
pSymborder.CornerRounding = 0;
IBorder pBorder = pSymborder;
pMapFrame.Border = pBorder;
pMapFrame.ExtentType = esriExtentTypeEnum.esriExtentBounds;
IElement pElement = (IElement)pMapFrame;
IEnvelope pEnvelop = new EnvelopeClass();
pEnvelop.PutCoords(0.5, 0.5, 29.2, 20.5); //这里设置 mapframe 的大小
IGeometry pGeometry = pEnvelop;
pElement.Geometry = pGeometry;
IPage pPage = axPageLayoutControl1.Page;
pPage.Orientation = 2;
pPage.PutCustomSize(29.7,21.0); //这里设置 page 的大小
axPageLayoutControl1.ActiveView.Refresh();
```

2）MapSurroundFrame 对象

MapSurroundFrame 是一种用于管理 MapSurround 对象的框架元素。MapSurround 是指北针、比例尺和图例一类的对象，它们是一种"智能"的，会自动与某个地图对象关联，随着地图视图的变化而变化的对象。当地图框架发生旋转的时候，指北针对象的方向也会发生变化。MapSurroundFrame 支持 MapFrameResized 事件，当地图的尺寸改变的时候，它会监听这个事件，并自动更新比例尺等对象。IMapSurroundFrame 接口是 MapSurroundFrame 对象的默认接口，IMapSurroundFrame 的 MapFrame 属性可以得到与自身关联的 MapFrame 对象，而 IMapSurroundFrame 的 MapSurround 属性则可以得到 Frame 中的 MapSurround 对象。

3）TableFrame 对象

EditorExt 库的 TableFrame 是一种可以容纳 Table 对象的框架元素，只能放置在布局视图中，而不能添加到 Map 对象中去。TableFrame 对应于 ArcMap 打开一个要素图层的属性表，点击"Table Options"按钮，选择"Add Table to Layout"后，会在布局视图中发现一份数据表。

由 ArcMapUI 库定义的 ITableFrame 接口是 TableFrame 类实现的唯一接口，它提供了操作框架元素中表的属性和方法。StartCol 和 StartRow 属性可以设置显示表时的列数和行数，TableView 属性可以得到一个 ITabelView 对象去改变表视图的属性，Table 属性可以返回给用户一个与框架相关的 ITable 对象，TableProperty 属性可以得到 ITableProperty 对象。

3. MapGrid 对象模型

地图格网在指示地理位置方面具有重要作用，在小比例尺地图中，经纬网可以指明某个区域在地球上的确切位置；在大比例地图里，也可以使用方格网将一块区域进行规则划分。

MapGrid 对象是布局视图中的一种参考线或点，由 Carto 库定义，可以帮助用户快速地确定地图中要素的位置。在 PageLayout 视图中，它一般出现在地图边缘，用于显示经纬度或者方格网。地理网格主要由 GridLine（格网线）、GridLabel（格网标注）和 GridBorder（格网边框）三部分组成。

MapGrid 是一个抽象类，它有五个子类对象：MeasuredGrid、Graticule、MgrsGrid、IndexGrid 和 CustomGridOverlay。MapGrid 组成了 MapFrame。IMapGrid 接口被所有的地图格网类继承，用于设置 MapGrid 对象的属性和方法。Border 属性用于设置地图网格的边框，LabelFormat 属性用于设置地图网格上的标签格式，LineSymbol 属性用于设置网线的样式，SetSubTicksVisiblity 方法可以设置 tick 的可见性。

Graticule 是使用经纬线来划分地图的地图格网对象，它实现了两个接口 IGraticule 和 IMeasuredGrid。由于 Graticule 对象使用经纬网，因而需要设置空间参考属性。IMeasuredGrid 接口定义了六个属性用于设置原点：FixedOrigin 属性为是否根据计算自动设置起点，Units 属性用于设置原点间隔的单位，XOrigin 和 YOrigin 属性用于设置 X、Y 方向上的起点，XIntervalSize 和 YIntervalSize 属性用于确定 X、Y 方向上两线之间的间隔。

MeasuredGrid 也是使用经纬度作为地图格网来划分地图的，它与 Graticule 对象的不同之处在于它的空间参考属性可以和 MapFrame 对象一致或不一致。它除了实现 IMeasureGrid 接口外，还实现了 IProjectedGrid 接口用于设置它的投影属性。

IndexGrid 是使用索引值的方式来划分地图的区域的对象，通常南北方向用"A、B、C、…"，东西方向用"1、2、3、…"来表示，它适合小区域内地块的划分。IndexGrid 类实现了 IIndexGrid 接口，XLabel 和 YLabel 属性分别用于设置网格 X、Y 轴上的标签，ColumnCount 和 RowCount 属性分别用于设置 MapGrid 网格划分的列数和行数。

下面是一个 IndexGrid 的例子：

```
private IIndexGrid CreateIndexGrid()
{
    IIndexGrid pIndexGrid = new IndexGridClass();
    pIndexGrid.ColumnCount = 5;
    pIndexGrid.RowCount = 5;
    int i;
    for (i = 0;i<= pIndexGrid.ColumnCount - 1;i++)
    {   pIndexGrid.set_XLabel(i,(i + 1).ToString()); }
    for (i = 0;i<= pIndexGrid.RowCount - 1;i++)
    {   pIndexGrid.set_YLabel(i, i.ToString () + "A");   }
    return pIndexGrid;
}
```

1）MapGridBorder 对象

地图格网的边框有两种类型：SimpleMapGridBorder 和 CalibratedMapGridBorder。它们都

实现了 IMapGridBorder 接口。DisplayName 属性可以得到边框的显示名，这两类边框的 DisplayName 分别是"simple border"和"calibrated border"。

SimpleMapGridBorder 对象只是使用简单的直线来作为地图的边框，因而在 ISimpleMapGridBorder 接口中必须设置的是 LineSymbol 属性，它用于确定边框线的样式、宽度和颜色。

CalibratedMapGridBorder 是使用一种渐变线段的边框对象，这个对象支持的接口是 ICalibratedMapGridBorder，它定义了边框的前景色、后景色、宽度、线段的间隔长度等属性。

2）MapGridLabel 对象

MapGridLabel 是用来设置地图格网的标签。由抽象类 GridLabel 定义，IGridLabel2 接口控制着所有 GridLabel 对象的一般属性。LabelAlignment 属性可以设置格网标注在格网对象的四个边上的水平和垂直方向，它需要传入一个 esriGridAxisEnum 枚举类型值。继承抽象类 GridLabel 的有 DMSGridLabel、FormattedGridLabel、MixedFontGridLabel、IndexGridLabel 和 CornerGridLabel 等。

DMSGridLabel 对象的特点是其标注字符使用的是经纬度的单位，即度、分和秒。DMSGridLabel 类实现 IDMSGridLabel3 接口用于管理经纬网标注对象的属性，如字体、标注类型等。

FormattedGridLabel 对象可以让标注上的字符格式化显示，实现了 IFormattedGridLabel 接口，其 Format 属性需要传入一个 INumberFormat 对象用于设定字符格式，如字符是否显示负号、小数点后设置多少位等。

MixedFontGridLabel 可以使用两种字体来设置一段标注文本，NumberOfDigits 属性可以确定两种字体是如何应用到标注字符上的，其中的 n 值可以让标签最后的 n 个字符设置为第二种字体和颜色，而剩下的字符使用第一种字体和颜色。MixedFontGridLabel 对象的主要颜色和字体使用 IGridLabel2 的 Color 和 Font 属性设置，第二种颜色和字体则由 IMixedFontGridLabel 接口的 SecondaryColor 和 SecondaryFont 属性确定。

3）MapGridFactory 对象

CartoUI 库定义的 MapGridFactory 对象可以让程序员快速新建一个地图格网对象，这些新建的格网对象的属性被设置为缺省值。MapGridFactory 是一个抽象类，它唯一的接口是 IMapGridFactory，其中定义的 Create 方法可以新建 MapGrid 对象。MapGridFactory 的子类有 GraticuleFactory、IndexGridFactory、MeasuredGridFactory、CustomOverlayGridFactory 和 MgrsGridFactory 五种。它们没有自己的接口，全部都是实现 IMapGridFactory 接口。

4. MapSurround 对象

MapSurround 是与一个地图对象关联的用于修饰地图的辅助图形元素对象。它们的形状或数值会随着地图属性的变化而自动改变，如指北针、比例尺等。MapSurround 是一个抽象类，它组成了 MapSurroundFrame 类。所有的 MapSurround 对象都支持 IMapSurround 接口，它定义了 MapSurround 对象一般的属性和方法，如使用 Name 属性可以得到某个 MapSurround 的名称，FitToBounds 方法可以改变一个 MapSurround 对象的尺寸。MapSurround 类也实现了 IMapSurroundEvents 接口，用于触发与 MapSurround 相关的事件，如 AfterDraw、BeforeDraw 等。

1）图例

图例（Legend）是与一个 Map 对象中图层的着色操作（Renderer）相关的对象，着色对象可以在地图上产生专题图。Legend 类的主要接口是 ILegend，使用它可以修改图例的属性和获得它的组成对象。Patch 是一个 LegendClass 中的帮助描述要素着色的图片，这个对象只有两种形式：AreaPatch 和 LinePatch。ArcObjects 使用 LegendClassFormat 和 LegendFormat 对象来管理一个图例项内的 Patch 对象。

每个地图的图例可以看作一个 LegendItem，每个 LegendItem 都有一个或者多个 Legend Group（图例组），而这个数目则取决于地图有多少种着色方案。每个 LegendGroup 都有一个或者多个 LegendClass（着色类）对象，而每个 LegendClass 代表了一个使用自身的符号和标签制作的图例分类。LegendItem 对象有多种类型：HorizontalBarLegendItem、Horizontal LegendItem、NestedLegendItem 和 VerticalLegendItem。

LegendClassFormat 对象用于控制单个 LegendItem 的外观，如 DescriptionSymbol、Line Patch 和 Patch 的属性等。LegendFormat 对象用于控制一个 Legend 的属性，特别是 Legend 内不同部分的间隔大小。

以下为插入图例代码：

```
IGraphicsContainer graphicsContainer = axPageLayoutControl1.GraphicsContainer;
IMapFrame mapFrame =
(IMapFrame)graphicsContainer.FindFrame(axPageLayoutControl1.ActiveView.FocusMap);
    if (mapFrame == null) return;
    UID uID = new UIDClass();
    uID.Value = "esriCarto.Legend";
    IMapSurroundFrame mapSurroundFrame = mapFrame.CreateSurroundFrame(uID, null);
    if (mapSurroundFrame == null) return;
    if (mapSurroundFrame.MapSurround == null) return;
    mapSurroundFrame.MapSurround.Name = "Legend";
    IEnvelope envelope = new EnvelopeClass();
    envelope.PutCoords(1, 1, 3.4, 2.4);
    IElement element = (IElement)mapSurroundFrame;
    element.Geometry = envelope;
    axPageLayoutControl1.AddElement(element, Type.Missing, Type.Missing, "Legend", 0);
    axPageLayoutControl1.ActiveView.PartialRefresh(esriViewDrawPhase.esriViewGraphics, null, null);
```

2）指北针

MarkerNorthArrow 是一种用于指示地图空间方位的图形，它其实是 ESRI North 字库中的字符符号，字库中的任何一种字体的符号都可以当作指北针使用。MarkerNorthArrow 从抽象类 NorthArrow 继承而来，是一个 MapSurround 对象。MarkerNorthArrow 对象的两个主要接口是 INorthArrow 和 IMarkerNorthArrow。INorthArrow 接口可以设置指北针对象的一般属性，如颜色、尺寸和引用位置；IMarkerNorthArrow 接口定义了一个属性 MarkerSymbol，它用于设置指北针的符号。

以下为添加指北针的代码，需要一个符号化控件进行交互。

```
GetSymbol symbolForm = new
```

```
GetSymbol(esriSymbologyStyleClass.esriStyleClassNorthArrows);
symbolForm.Text = "选择指北针";
IStyleGalleryItem styleGalleryItem =
symbolForm.GetItem(esriSymbologyStyleClass.esriStyleClassNorthArrows);
symbolForm.Dispose();
if (styleGalleryItem == null) return;
IMapFrame mapFrame =
(IMapFrame)m_hookHelper.ActiveView.GraphicsContainer.FindFrame(m_hookHelper.ActiveView.FocusMap);
IMapSurroundFrame mapSurroundFrame = new MapSurroundFrameClass();
mapSurroundFrame.MapFrame = mapFrame;
mapSurroundFrame.MapSurround = (IMapSurround)styleGalleryItem.Item;
IElement element = (IElement)mapSurroundFrame;
element.Geometry = envelope;
m_hookHelper.ActiveView.GraphicsContainer.AddElement((IElement)mapSurroundFrame, 0);
m_hookHelper.ActiveView.PartialRefresh(esriViewDrawPhase.esriViewGraphics, mapSurroundFrame, null);
```

3）比例尺

ScaleBar 对象也是一种 MapSurround，它有多个子类，如 ScaleLine、SinglefillScaleBar 和 DoublefillScaleBar 等，这些类都实现了 IScaleBar 和 IScaleMarks 接口。IScaleBar 接口可以管理一个比例尺对象的大部分属性，如比例尺颜色、高度，它也定义了管理比例尺对象上 Label 的属性，如 Labelsymbol、LabelPosition 等，它们分别用于设置比例尺中的标识字符符号和位置。IScaleMarks 接口负责管理与一个比例尺相关的单个标记（mark）的属性，如高度、符号和位置等。

SinglefillScaleBar 是个抽象类，它的对象使用一种符号来显示比例尺。这个类实现的接口是 ISinglefillScaleBar，其子类为 SingleDivisionScaleBar，子类没有自己特殊的接口。

DoublefillscaleBar 是使用两种符号来交叉显示比例尺的，其接口是 IDoublefillScaleBar，使用它的 FillSymbol1 和 FillSymbol2 属性可以设置 DoublefillScaleBar 中的两种符号。DoublefillScaleBar 有三个子类：AlternatingScaleBar、DoubleAlternatingScaleBar 和 Hollowscale Bar。

ScaleLine 对象是使用步进线（stepped-line）来显示比例尺的，它本身是一个组件类，实现了 IScaleLine 接口。SteppedScaleLine 对象是 ScaleLine 的子类。

4）文本比例尺

比例尺能够用图形的方式显示出地图上的单位长度在现实世界的距离，但用户一般都希望在地图上能够获得一个明确的比例值。ArcObjects 提供的比例文本对象 ScaleText 就可以满足这个要求。ScaleText 对象在本质上是一个文本元素，但是它会随着相关地图的变化而改变比例值。ScaleText 类实现了 IScaleText 接口，它定义了文本的格式，如 symbol、style 等。用户也可以通过 text 只读对象来得到比例文本的字符值。

以下为添加文本比例尺的实例：

```
IEnvelope pEnv;
pEnv = pPageLayoutControl.TrackRectangle();
GetSymbol symbolForm = new GetSymbol(esriSymbologyStyleClass.esriStyleClassScaleTexts);
symbolForm.Text = "选择比例尺文本";
IStyleGalleryItem styleGalleryItem =
symbolForm.GetItem(esriSymbologyStyleClass.esriStyleClassScaleTexts);
symbolForm.Dispose();
if (styleGalleryItem == null) return;
IMapFrame mapFrame =
(IMapFrame)m_hookHelper.ActiveView.GraphicsContainer.FindFrame(m_hookHelper.ActiveView.FocusMap);
IMapSurroundFrame mapSurroundFrame = new MapSurroundFrameClass();
mapSurroundFrame.MapFrame = mapFrame;
mapSurroundFrame.MapSurround = (IMapSurround)styleGalleryItem.Item;
IElement element = (IElement)mapSurroundFrame;
element.Geometry = pEnv;
m_hookHelper.ActiveView.GraphicsContainer.AddElement((IElement)mapSurroundFrame, 0);
m_hookHelper.ActiveView.PartialRefresh(esriViewDrawPhase.esriViewGraphics, mapSurroundFrame, null);
```

5. 地图输出

地图输出分为两种类型：打印输出和转换输出。打印输出即调用与计算机连接的打印设备将地图打印在纸质媒介上；转换输出即将地理数据输出为不同格式的文件，如 JPEG、PDF、BMP 等。Carto 库中的 PrintAndExport 类实现了 IPrintAndExport 接口，该接口的 Export、ExportPages 方法分别用于输出指定的视图、页面；Print、PrintPages 方法分别用于打印指定的视图、页面；PrinterNames 属性返回所有可用的打印机的名称；PageCount 属性返回数据驱动页面数；PageRow 属性返回指定的数据驱动页面的行。

在 ArcObjects 中，PrintAndExport 类是基于数据驱动页面输出地图的唯一方法。如果想用少量代码将 Map 或 PageLayout 的视图输出到一个文件，可以使用 PrintAndExport 类，这个类通常与输出类（如 ExportPNG、ExportPDF 等）一起使用产生图形输出文件，此时的图像重采样、输出影像的质量、可视范围的管理及其他复杂的处理都在内部完成。如果地图的输出需要更精细的控制，如需要计算 VisibleBounds、PixelBounds，或需要控制其他的输出及显示参数，那么 PrintAndExport 类不能再继续使用，需要开发人员编程实现更精细的控制。此时可以使用 IExport 接口中 StartExporting、FinishExporting 方法及 IActiveView 接口中的 Output 方法来实现。

1）地图的打印输出

地图的打印输出，除了使用 PrintAndExport 类外，还可以使用 Printer 相关的类。Output 库中的 Printer 类是一个抽象类，它有三个子类：EmfPrinter、ArcPressPrinter 和 PsPrinter。EmfPrinter 对象是通过使用 EMF 数据作为打印驱动，PsPrinter 对象是使用 PostScript 作为驱动来打印输出地图，ArcPressPrinter 则是使用 ArcPressPrinterDriver 来输出地图。

2）地图的转换输出

地图的转换输出，除了使用 PrintAndExport 类外，还可以使用 Export 相关的类。Output 库中的 Export 类是所有转换输出类的父类，是一个抽象类，其 IExport 接口用于定义转换输出对象的一般方法和属性。IExport 接口应结合输出类一起使用，共有十个输出类，分为两大类：①基于影像的 ExportImage：ExportBMP、ExportGIF、ExportJPEG、ExportPNG 和 ExportTIFF，它们都实现了 IExportImage 接口；②基于矢量的 ExportVector：ExportAI (Adobe Illustrator)、ExportEMF、ExportPDF、ExportPS 和 ExportSVG(Scalable Vector Graphics)，它们都实现了 IExportVector 接口。

自主练习：根据 Carto 和 Output 库中类和接口的定义，实现地图的整饰与输出。

第三部分　基于 GeoServer 的 WebGIS 开发系列实验

实验 3-1　GeoServer 的安装与使用

GeoServer 是 Java 语言编写的开源软件，可以用于地理数据的共享和编辑服务。GeoServer 兼容 Open Geospatial Consortium（OGC）的 Web Feature Service（WFS）和 Web Coverage Service（WCS）标准规范，同时支持 Web Map Service（WMS）高效率渲染。

GeoServer 支持 PostGIS、Shapefile、ArcSDE、Oracle、VPF、MySQL 等多种数据格式。GeoServer 支持上百种投影，能够将网络地图输出为 JPEG、GIF、PNG、SVG、KML 等格式。

本实验采用 GeoServer+Tomcat+Openlayer 进行 WebGIS 开发。

（1）实验目的：本实验是 WebGIS 原理系列实验的前期准备实验，侧重对相关实验环境的初步认识和对空间数据来源途径的基本了解，为后续实验操作和自主练习提供帮助。

（2）相关实验：GIS 专业实验设备与环境配置中的"GIS 应用开发环境"和"GIS 应用开发资源"。

（3）实验数据：本教材系列实验数据。

（4）实验环境：GeoServer、Java JDK。

（5）实验内容：安装 GeoServer；配置开发环境。

1. JDK 安装与环境配置

1）安装 JDK

可从 Oracle 官方网站下载 Java 最新版本。JDK 下载地址：https://www.oracle.com/technetwork/java/javase/downloads/index.html。

下载完毕后，直接运行 JDK 安装软件，默认安装即可。

打开软件安装程序，弹出 JDK 安装程序启动界面，点击"下一步"，打开安装目标文件夹，可对目标路径进行更改，点击"确定"，即开始安装，安装完成后点击关闭。

2）配置 JDK 环境变量

打开"我的电脑"，选择系统属性，点击"高级系统设置"，选择环境变量，打开环境变量窗口。环境变量窗口如图 3-1-1 所示。

在环境变量窗口中，选择系统变量→新建，即可新建一个变量。新建变量名：JAVA_HOME，变量值：C:\Program Files\Java\jdk-12.0.2，即之前安装 JDK 的目标路径，读者可根据自己的安装情况更改，如图 3-1-2 所示。

在系统变量中，选择 Path 变量，对变量进行编辑。如图 3-1-3 所示，选择编辑 Path 系统变量，增添变量值：%JAVA_HOME%\bin、%JAVA_HOME%\jre\bin。

在系统变量中接着新建一个变量。新建系统变量名：CLASSPATH，变量值：.;%JAVA_HOME%\lib\dt.jar;%JAVA_HOME%\lib\tools.jar，如图 3-1-4 所示。

图 3-1-1　环境变量窗口

图 3-1-2　新建变量示意图

图 3-1-3　编辑 Path 变量示意图

图 3-1-4　新建 CLASSPATH 变量示意图

3）检测 JDK 安装情况

同时按下"Win"+"R"，打开命令框，输入"CMD"，在"CMD"命令下输入"javac"命令，出现如图 3-1-5 所示界面，表示安装成功。

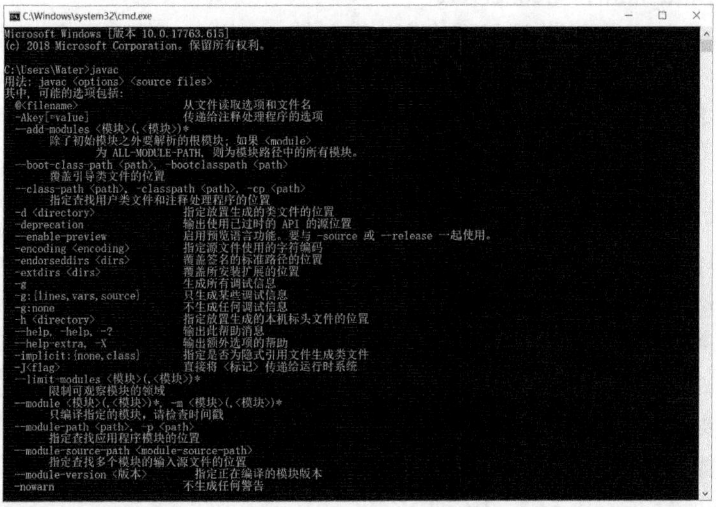

图 3-1-5　javac 命令界面

2. GeoServer 的安装与运行

1）下载文件

下载地址：http://GeoServer.org/download/或 http://sourceforge.net/projects/GeoServer。

下载文件名称　GeoServer-a.b.c-bin.zip，其中 a、b、c 为版本号。本实验示例文件为 GeoServer-2.15.2-bin.zip。将文件解压缩，例如，解压到 C:\GeoServer，建立一个系统环境变量，变量名为 GeoServer_HOME，变量值为解压缩的地址，如本实验中的 C:\GeoServer。

2）安装 GeoServer

在 Windows 命令窗口输入：%GeoServer_HOME%\bin\startup。

按照提示，安装完毕后，在浏览器中输入：http://localhost:8080/GeoServer/web/。检查是否能看到默认的欢迎界面。如果不能，请检查 8080 端口是否被其他服务占用。检查方法如下，如图 3-1-6 所示，在命令窗口中，输入 netstat -ano|findstr "8080"，检查到 8080 端口被"14300"占用；输入命令：tasklist|findstr "14300"，检查到该端口被 360se.exe 所占用。

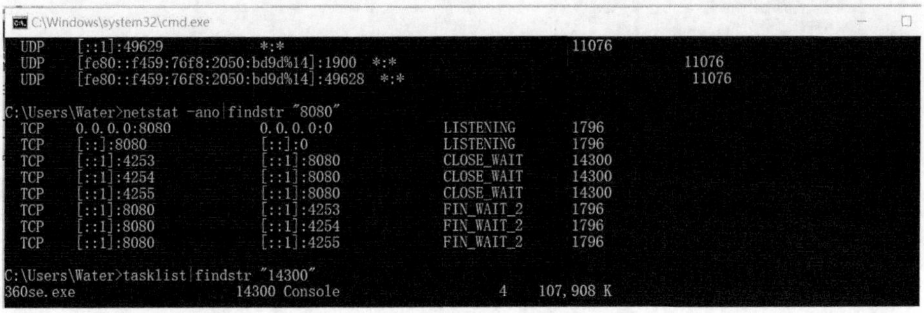

图 3-1-6　检查 8080 端口是否被占用

如被占用，可修改$GeoServer_HOME/start.ini 文件中的 8080，修改为可用的端口名并保存，欢迎界面修改为 8090。相应地，在浏览器中输入：http://localhost:####/GeoServer/web/。

在 GeoServer 欢迎界面的右上角，有账户和密码输入框与登录按钮。GeoServer 默认的用户名和密码分别是 admin 和 GeoServer。

登录成功后，GeoServer 欢迎界面将变为默认的 GeoServer 管理界面，操作和配置修改的选项都在界面的左边，以链接的形式给出。可以通过点击不同的超链接进入相应的管理界面。

登录之后，可以在 Security 选项下对用户、密码、管理组、权限角色等进行修改和配置。

实验 3-2　地图图层的发布与管理

本实验的前提条件是成功搭建 GeoServer 环境。实验内容为基于 GeoServer 部署地图数据。对于 GeoServer 应用系统来说有三大重要知识点，分别为工作区（又称工作空间，workspace）、数据存储器（store）和地图图层（layer），这些概念都将在本实验中逐一介绍。

（1）实验目的：了解 GeoServer 的基本构成，学会使用 GeoServer 来发布本地地图。

（2）相关实验：实验 3-1 GeoServer 的安装与使用。

（3）实验数据：GeoServer 共享数据。

（4）实验环境：GeoServer、Java JDK。

（5）实验内容：数据准备；新建工作区；新建数据存储器；添加地图图层和图层组；发布地图图层；地图预览。

1. 数据准备

下载地图数据 nyc_roads.zip，shp 文件格式，为纽约市路网数据。下载地址为：http://docs.geoserver.org/2.3.3/user/_downloads/nyc_roads.zip。

将文件解压到%GeoServer_HOME%/data_dir/data/nyc_roads 目录。实验 3-1 示例中，GeoServer_HOME 的路径为 C:\geoserver，因此本实验中将地图数据解压至 C:\geoserver/data_dir/data/nyc_roads。

2. 新建工作区

工作区是组织管理地图图层的容器。通常，每个项目对应一个工作区。

启动浏览器，登录 GeoServer，进入 GeoServer 管理界面。点击左侧"工作区"，可看到当前系统中已建的工作区，如 cite、sf 等，如图 3-2-1 所示。

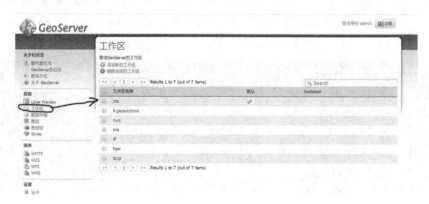

图 3-2-1　GeoServer 工作区浏览界面

在工作区浏览界面中，点击"添加新的工作区"，工作区以英文字母命名，名称长度小于 10 个字母，且中间不能有空格。如图 3-2-2 所示，新建了一个名称为 nyc 的工作区，命名空间 URI 为 http://geoserver.org/nyc。此处的命名空间将在查找 WFS 时被浏览到。

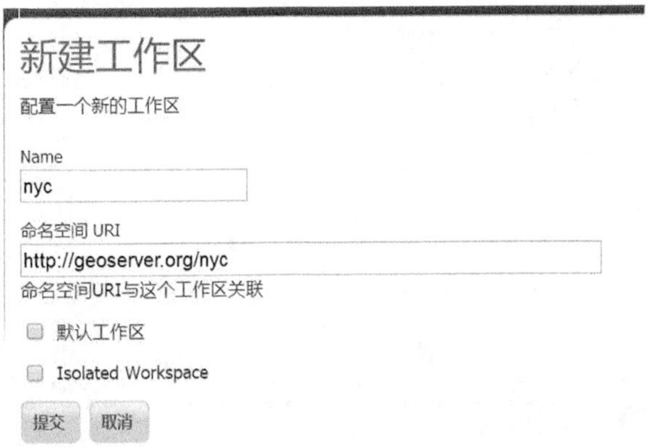

图 3-2-2　GeoServer 新建工作区界面

3. 新建数据存储器

在图 3-2-3 的数据存储器浏览界面中，可"添加新的数据存储"。

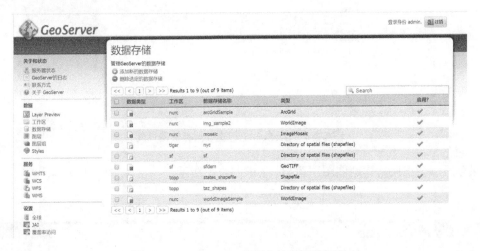

图 3-2-3　GeoServer 数据存储器浏览界面

在添加的页面中，选择矢量数据源中的 Shapefile，如图 3-2-4 所示。Shapefile 为矢量数据。因此，在新建矢量数据界面，输入工作区的名称、数据源的名称、Shapefile 的存放路径等。此处一定要注意填写数据源的名称和选择工作区，其中数据源名称是指要添加的地图图层名，工作区是指需要把添加的矢量数据放到哪个工作区。本实验选择的是刚刚建立的 nyc 工作区，数据源名称为 nyc_roads。

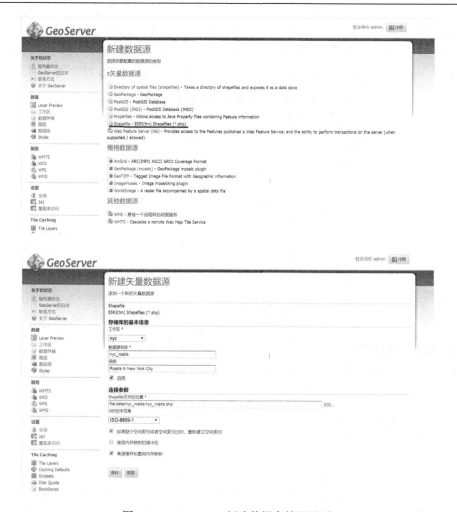

图 3-2-4　GeoServer 新建数据存储器界面

保存后，会出现新建地图图层窗口，如图 3-2-5 所示。

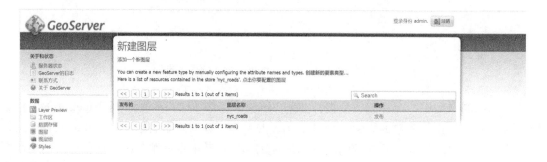

图 3-2-5　新建地图图层界面

4. 发布地图图层

在图 3-2-5 相应图层操作中点击"发布"，进入编辑地图图层界面，如图 3-2-6 所示。

图 3-2-6　地图图层编辑界面

在编辑图层的界面中，可以定义数据范围，也可以根据地图数据本身的坐标提取边界（Compute from data、Compute from native bounds）。选择要发布的地图图层数据的坐标系，然后计算地图数据的坐标范围，点击保存，如图 3-2-7 所示。

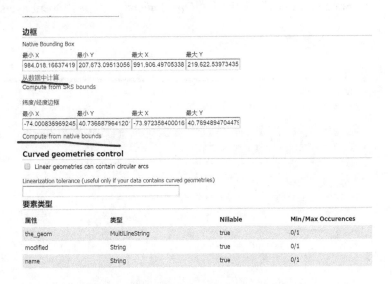

图 3-2-7　地图图层坐标系统和范围编辑界面

在发布地图图层界面中，要确保默认的地图类型与数据类型一致。本实验地图数据是线矢量数据，如图 3-2-8 所示，因此需要将 Default Style 设置为 line。

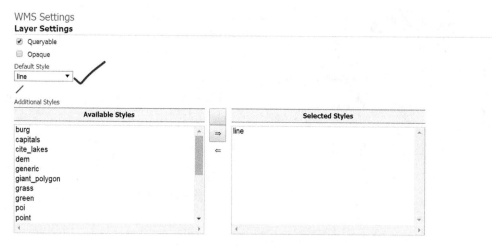

图 3-2-8　地图图层发布类型编辑界面

5. 地图预览

接下来，通过地图预览验证刚才发布的地图是否有效。如图 3-2-9 所示，在管理界面的左侧选择"Layer Preview"图层预览，进入地图图层预览管理界面。搜索找到刚才发布的地图图层，选择"Common Formats"列的中"OpenLayers"，点击"OpenLayers"则可访问发布的地图。

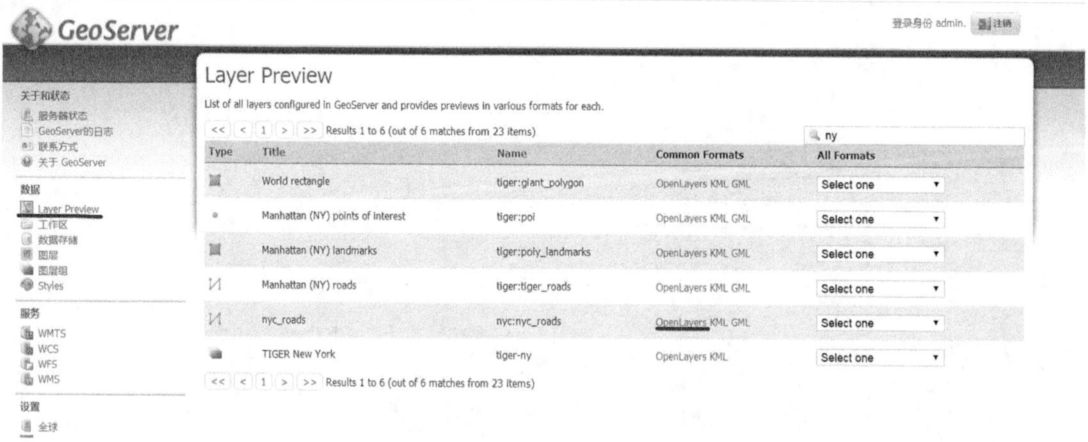

图 3-2-9　地图图层预览界面

本实验以刚刚建立的图层 nyc_roads 为例，点击该地图后面的 OpenLayers，即可访问刚才发布的地图，结果如图 3-2-10 所示。

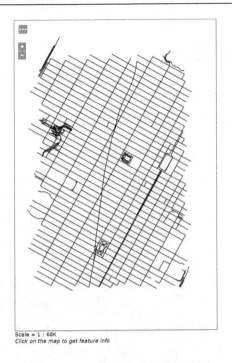

图 3-2-10　nyc_roads 地图图层预览

自主练习：添加点要素、面要素数据，并发布地图。

实验 3-3　地　图　浏　览

本实验采用 HBuilderX 提升开发效率。HBuilderX 支持 HTML5 的 Web 开发。本实验练习采用实验 3-2 中所发布的地图。OpenLayers 是用于 WebGIS 客户端的 JavaScript 包，支持 Open GIS 协会制定的 WMS（Web Mapping Service）和 WFS（Web Feature Service）等网络服务规范，支持地图图层叠加显示。

（1）实验目的：通过实验，初步认识 HBuilder 的基本架构，掌握 HBuilder 的主要功能和特性，实现对地图视图的调用和浏览操作。

（2）相关实验：实验 3-2 地图图层的发布与管理。

（3）实验数据：OSM 在线数据资源、GeoServer 示例数据。

（4）实验环境：GeoServer、Java JDK、HBuilderX。

（5）实验内容：①使用 Map 控件和代码编写一个窗体应用程序实现加载网络地图的功能；②使用 OpenLayers 控件和代码编写一个窗体应用程序实现对地图的放大、缩小、漫游等基本视图操作功能。

1. HBuilderX 基本操作

HBuilderX 是绿色发行包（不含插件），下载解压后，直接运行 HBuilderX 应用程序即可。HBuilderX 支持 java 插件、nodejs 插件，并可兼容很多 vscode 的插件及代码块。

打开 HBuilderX，界面如图 3-3-1 所示，点击文件菜单，选择新建项目。打开项目创建窗口，如图 3-3-2 所示，填写项目名称，设置保存位置，点击"创建"即可。

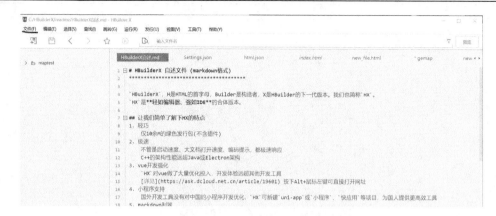

图 3-3-1　HBuilder X 界面

图 3-3-2　创建项目界面

创建好项目后，在左侧就出现了刚创建的项目名称，本实验示例项目名为 maptest。选中项目名，右击新建 html 文件。打开 html 创建窗口，设置名称和存储位置，点击"创建"即可，如图 3-3-3 所示。创建完成 html 文件后，就可以在 HBuilder 中进行编码操作。

图 3-3-3　新建 html 文件界面

2. OpenLayers 简介

1）Map

Map（ol.Map）是 OpenLayers 的核心部件。Map 呈现到对象 target 容器。

例如，网页上的 div 元素作为 Map 容器，示例代码如下：

```
<div id="my_map" style="width: 100%, height: 400px"></div>
<script>
var map = new ol.Map({target: 'my_map'});
</script>
```

2）View

View（ol.View）负责对地图的浏览设置，包括地图的投影（projection）、中心点（center）、缩放比例（zoom）等。默认的投影是墨卡托投影（EPSG：3857），默认的地图单位是米。

地图的缩放级别由 maxZoom （默认值为 28）、zoomFactor （默认值为 2）、maxResolution 确定。

设置 View 示例如下：

```
map.setView(new ol.View({
center: [0, 0],
zoom: 2
}));
```

3）Source

Source（ol.source.Source）负责获取远程数据图层，包含免费的和商业的在线地图，如 OpenStreetMap、Bing、OGC 资源（WMS 或 WMTS）、矢量数据（GeoJSON 格式、KML 格式等）等。所需代码为：

```
var osmSource = new ol.source.OSM();
```

4）Layer

Layer 是地图图层，OpenLayers 3 包含三种基本图层类型：ol.layer.Tile、ol.layer.Image 和 ol.layer.Vector。

ol.layer.Tile 用于显示瓦片资源。

ol.layer.Image 用于显示支持渲染服务的影像图片。

ol.layer.Vector 用于显示渲染矢量数据资源。

添加瓦片资源地图图层示例：

```
var osmLayer = new ol.layer.Tile({source: osmSource});
map.addLayer(osmLayer);
```

3. 显示地图

OpenStreetMap (OSM)提供了免费的地图数据服务，可以在 OpenLayers 中作为地图数据源使用。

首先构造一个地图容器`<div id="my_map"></div>`，然后创建 Map，添加地图图层，设置渲染展示采用的地图投影、中心点、缩放比例等，最后在浏览器测试。

参考示例代码如下：

```
<!doctype html>
```

```html
<html lang="en">
<head>
    <link rel="stylesheet" href="http://localhost:8090/geoserver/openlayers3/ol.css" type="text/css">
    <title>OpenLayers 3 example</title>
    <script src="http://localhost:8090/geoserver/openlayers3/ol.js" type="text/javascript"></script>
    <style>
    #my_map {
    width: 512px;
    height: 256px;
    }
    .ol-attribution a {
    color: black;
    }
    </style>
</head>
<body>
    <h1>My Map</h1>
    <div id="my_map"></div>
        <script>
        var map = new ol.Map({
        target: 'my_map',
        layers: [
        new ol.layer.Tile({
        source: new ol.source.OSM()
        })
        ],
        view: new ol.View({
        center: ol.proj.transform([-93.27, 44.98], 'EPSG:4326', 'EPSG:3857'),
        zoom: 9
        }),
        controls: ol.control.defaults({
        attributionOptions: {
        collapsible: false
        }
        })
        });
        </script>
```

    </body>
</html>

编写完成后，代码保存为 html 文件，在 HBuilderX 界面点击"运行"菜单，选择相应浏览器，就可以显示出所发布的地图。本实验示例加载 OSM 地图数据，测试结果如图 3-3-4 所示。

图 3-3-4 加载 OSM 地图数据测试结果

自主练习：编写 html 文件，实现对自己已发布的地图的加载显示。

提示：可在实验 3-2 发布地图后，在地图预览结果页面，鼠标右击查看源码，学习 map 的创建、layer 的添加、view 的设置等。

添加 GeoServer 发布的 nyc_roads 图层的参考示例代码如下：

<!doctype html>
<html lang="en">
<head>
<link rel="stylesheet" href="http://localhost:8090/geoserver/openlayers3/ol.css" type=" text/css">
    <style>
        #map {
            clear: both;
            position: relative;
            width: 506px;
            height: 768px;
            border: 1px solid black;
        }
    </style>
    <title>OpenLayers vector map example</title>
<script src="http://localhost:8090/geoserver/openlayers3/ol.js" type="text/javascript">
</script>

```html
</head>
<body>
    <h1>My Map</h1>
    <div id="map"></div>
    <script type="text/javascript">
        var projection = new ol.proj.Projection({
            code: 'EPSG:2908',
            units: 'm',
            global: false
        });
        var bounds = [984018.1663741902, 207673.09513056703,
            991906.4970533887, 219622.53973435296
        ];
        var map = new ol.Map({
            target: 'map',
            layers: [
                new ol.layer.Tile({
                title: "New York City",
                source: new ol.source.TileWMS({
                url: 'http://localhost:8090/geoserver/nyc/wms',
                params: {
                    'FORMAT': 'image/png',
                    'VERSION': '1.1.1',
                    tiled: true,
                    "LAYERS": 'nyc:nyc_roads',
                    "exceptions": 'application/vnd.ogc.se_inimage',
                    tilesOrigin: 984018.1663741902 + "," + 207673.09513056703
                    }
                })
                })
            ],
            view: new ol.View({
                projection: projection,
            })
        });
        map.getView().fit(bounds, map.getSize());
    </script>
</body>
</html>
```

4. 地图浏览控件

创建比例尺控件 ScaleLine，该控件用于显示当前地图窗口的地图比例尺。在创建地图的代码中，添加如下代码，为地图创建默认的比例尺控件：

```
controls: ol.control.defaults().extend([
new ol.control.ScaleLine()
]),
```

例如，在实验 3-2 地图显示的代码中，添加比例尺控件，则在地图视图的左下角显示默认的比例尺。

含有比例尺控件的地图显示的参考代码为：

```
var map = new ol.Map({
            target: 'map',
            layers: [
                new ol.layer.Tile({
                    title: "New York City",
                    source: new ol.source.TileWMS({
                        url: 'http://localhost:8090/geoserver/nyc/wms',
                        params: {
                            'FORMAT': 'image/png',
                            'VERSION': '1.1.1',
                            tiled: true,
                            "LAYERS": 'nyc:nyc_roads',
                            "exceptions": 'application/vnd.ogc.se_inimage',
                            tilesOrigin: 984018.1663741902 + "," +
                            207673.09513056703
                        }
                    })
                })
            ],
            //添加地图比例尺控件
            controls: ol.control.defaults().extend([
            new ol.control.ScaleLine()
            ]),
            view: new ol.View({
                projection: projection,
            })
        });
```

运行以上代码，其结果如图 3-3-5 所示。在显示发布地图的基础上，左下角添加了比例尺控件。

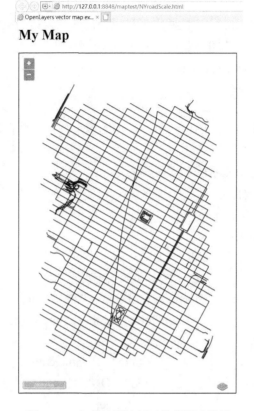

图 3-3-5　加载地图比例尺控件测试结果

自主练习：练习添加以下地图控件，

ol.control.FullScreen()——全屏组件，控件默认显示在地图视图的右上角。

ol.control.MousePosition()——鼠标位置控件，控件默认在地图视图的右上角显示鼠标所处的坐标。

ol.control.OverviewMap()——显示概略图控件，控件默认显示在地图视图的左下角。

ol.control.ZoomSlider()——放大、缩小控件，控件默认显示在地图视图的左边框。

ol.control.ZoomToExtent()——缩放到指定范围控件，地图缩放至确定的坐标范围。

实验 3-4　地图图层叠加显示

本实验练习采用 OpenLayers 显示多个地图图层。通过叠加图层，以及叠加圆、点等图形，初步了解应用 OpenLayers 进行叠加图层的流程。

（1）实验目的：通过实验，认识地图图层类型和特性，实现对地图图层的叠加显示和浏览操作。

（2）相关实验：实验 3-3　地图浏览。

（3）实验数据：OSM 在线数据资源、GeoServer 示例数据。

（4）实验环境：GeoServer、Java JDK、HBuilderX。

（5）实验内容：叠加图层，实现多图层的可视化浏览；同时添加圆与点图形，实现动态图层的渲染。

1. 图层叠加显示

多个图层可以添加到 layers 中，格式如下：

```
layers: [
   layer1,
     layer2
  ],
```

地图渲染时，将按照先绘制 layer1，再绘制 layer2 的顺序进行展示。

例如，创建两个图层，分别是 OSM 地图图层 osmLayer 和点图层 pointLayer，其示例代码为：

```
var osmLayer = new ol.layer.Tile({
    source: new ol.source.OSM()
});
var pointLayer = new ol.layer.Vector({
    source: new ol.source.Vector()
});
```

然后，将这两个图层添加到 map 中，代码为：

```
new ol.Map({
    // 在地图上添加上面创建的两个图层，图层顺序自下而上依次是 osmLayer、pointLayer
    layers: [osmLayer, pointLayer],
    view: new ol.View({
        center: [0, 0],
        zoom: 2
    }),
    target: 'map'
});
```

自主练习：多图层叠加显示。

2. 叠加图形

创建一个点图层，向图层中添加一个点，点的坐标为（0,0），设置点的显示样式为绿色填充、边线为红色的圆点。

示例代码如下：

```
//创建一个点图形图层
 var pointLayer = new ol.layer.Vector({
    source: new ol.source.Vector()
 });
//创建一个点
 var point = new ol.Feature({
```

```
                geometry: new ol.geom.Point([0, 0])
        });
        point.setStyle(new ol.style.Style({    //设置点的显示样式
                image: new ol.style.Circle({
                    radius: 7,
                    fill: new ol.style.Fill({
                        color: 'green'      //设置点符号的填充颜色
                    }),
                    stroke: new ol.style.Stroke({ //设置点的边线样式
                        color: 'red',
                        size: 10
                    })
                })
        }));
        pointLayer.getSource().addFeature(point);   //将矢量 point 图形添加到已创建的图层 pointLayer。
```

自主练习：创建图层，添加一个圆形图形，通过调整图层叠加顺序，查看显示效果。

参考示例代码如下：

```
<!Doctype html>
<html xmlns=http://www.w3.org/1999/xhtml>
<head>
    <meta http-equiv=Content-Type content="text/html;charset=utf-8">
    <meta http-equiv=X-UA-Compatible content="IE=edge,chrome=1">
    <meta content=always name=referrer>
    <title>叠加图形</title>
    <link rel="stylesheet" href="http://localhost:8090/geoserver/openlayers3/ol.css" type="text/css">
    <script src="http://localhost:8090/geoserver/openlayers3/ol.js" type="text/javascript"></script>
</head>
<body>
<div id="map" style="width: 100%"></div>
<div> 显示/隐藏：
    <input type="checkbox" checked="checked" onclick="checkOsm(this);" />OSM 地图
    <input type="checkbox" checked="checked" onclick="checkCircle(this);"/>圆形
    <input type="checkbox" checked="checked" onclick="checkPoint(this);"/>点
</div>
<div>
    图层顺序：
```

```html
        <input name="seq" type="radio" value="" onclick="upOsm(this);" />地图最上
        <input name="seq" type="radio" value="" checked="checked" onclick="upCircle(this);"/>圆形最上
        <input name="seq" type="radio" value="" onclick="upPoint(this);"/>点最上
    </div>
    <script>
```
```javascript
        // 创建三个图层，分别是OSM地图图层、点图层、圆形图层
        var osmLayer = new ol.layer.Tile({
            source: new ol.source.OSM()
        });
        var pointLayer = new ol.layer.Vector({
            source: new ol.source.Vector()
        });
        var circleLayer = new ol.layer.Vector({
            source: new ol.source.Vector()
        });

        new ol.Map({
            // 在地图上添加上面创建的三个图层
            layers: [osmLayer, pointLayer, circleLayer],
            view: new ol.View({
                center: [0, 0],
                zoom: 2
            }),
            target: 'map'
        });

        // 添加点，坐标为(0,0)
        var point = new ol.Feature({
            geometry: new ol.geom.Point([0, 0])
        });
        point.setStyle(new ol.style.Style({   //设置点符号样式
            image: new ol.style.Circle({
                radius: 7,
                fill: new ol.style.Fill({
                    color: 'green'
                }),
                stroke: new ol.style.Stroke({ //点符号的边线样式
                    color: 'red',
```

```
                    size: 10
                })
            })
        }));
        pointLayer.getSource().addFeature(point);   //添加矢量 point 到 pointLayer

        // 添加圆图形
        var circle = new ol.Feature({
            geometry: new ol.geom.Point([0, 0])
        });
        circle.setStyle(new ol.style.Style({
            image: new ol.style.Circle({
                radius: 10,
                stroke: new ol.style.Stroke({
                    color: 'blue',
                    size: 1
                })
            })
        }));
        circleLayer.getSource().addFeature(circle);

        // 隐藏显示 osmLayer
        function checkOsm(elem) {
            osmLayer.setVisible(elem.checked);
        }
        // 隐藏显示 pointLayer
        function checkPoint(elem) {
            pointLayer.setVisible(elem.checked);
        }
        // 隐藏显示 circleLayer
        function checkCircle(elem) {
            circleLayer.setVisible(elem.checked);
        }
        // osmLayer 图层到最上层显示
        function upOsm (elem) {
            if (elem.checked) {
                osmLayer.setZIndex(3);
                circleLayer.setZIndex(circleLayer.getZIndex()-1);
                pointLayer.setZIndex(pointLayer.getZIndex()-1);
```

```
            }
        }
        // circleLayer 图层到最上层显示
        function upCircle (elem) {
            if (elem.checked) {
                circleLayer.setZIndex(3);
                osmLayer.setZIndex(osmLayer.getZIndex()-1);
                pointLayer.setZIndex(pointLayer.getZIndex()-1);
            }
        }
        // pointLayer 图层到最上层显示
        function upPoint(elem) {
            if (elem.checked) {
                pointLayer.setZIndex(3);
                osmLayer.setZIndex(osmLayer.getZIndex()-1);
                circleLayer.setZIndex(circleLayer.getZIndex()-1);
            }
        }
    </script>
</body>
</html>
```

3. 叠加 GeoJSON 数据

GeoJSON 是一种基于 Javascript 对象表示地理数据的数据编码格式。GeoJSON 支持的几何类型包括点、线、面、多点、多线、多面和几何集合。GeoJSON 对象的坐标参考系统（coordinate reference system, CRS）是由 CRS 对象来确定的。

1）点要素

点由一个坐标对组成，格式如下：

```
{
    "type": "Point",
    "coordinates": [10.0, 10.0]
}
```

2）线要素

线由两个坐标组成，格式如下：

```
{
    "type": "LineString",
    "coordinates":
    [
        [20.0, 0.0],
        [41.0, 10.0]
```

        ]
    }

3）面要素

面由首尾相同的坐标序列组成，格式如下：

{
    "type": "Polygon",
    "coordinates":
    [
        [
            [10.0, 0.0],
            [19.0, 0.0],
            [19.0, 1.0],
            [10.0, 0.0]
        ]
    ]
}

4）多元素集合

多点、多线、多面以相对应图形元素坐标序列组成，元素之间以逗号隔开。以多点为例，代码如下：

{
    "type": "MultiPoint",
    "coordinates":
    [
        [50.0, 0.0],
        [11.0, 11.0]
    ]
}

几何集合中的每个元素都是上面所描述的几何对象之一，例如：

{
    "type": "GeometryCollection",
    "geometries":
    [
        {
            "type": "Point",
            "coordinates": [0.0, 0.0]
        },
        {
            "type": "LineString",
            "coordinates":

```
            [
                [1.0, 0.0],
                [7.0, 1.0]
            ]
        }
    ]
}
```
自主练习：创建 GeoJSON 数据源，添加点、线、面覆盖物标记，查看显示效果。

叠加 GeoJSON 数据应用示例代码如下：

```html
<!DOCTYPE html>
<html>
<head>
    <title>添加点、线、面覆盖物标记</title>
    <link rel="stylesheet" href="http://localhost:8090/geoserver/openlayers3/ol.css" type="text/css">
    <script src="http://localhost:8090/geoserver/openlayers3/ol.js" type=" text/javascript"> </script>
</head>
<body>
<div id="map" class="map"></div>
<script>
    var image = new ol.style.Circle({
        radius: 3,
        fill: null,
        stroke: new ol.style.Stroke({color: 'red', width: 1})
    });

    var styles = {
        'Point': new ol.style.Style({
            image: image
        }),
        'LineString': new ol.style.Style({
            stroke: new ol.style.Stroke({
                color: 'green',
                width: 1
            })
        }),
        'MultiLineString': new ol.style.Style({
            stroke: new ol.style.Stroke({
```

```
                    color: 'green',
                    width: 1
                })
            }),
            'MultiPoint': new ol.style.Style({
                image: image
            }),
            'MultiPolygon': new ol.style.Style({
                stroke: new ol.style.Stroke({
                    color: 'yellow',
                    width: 1
                }),
                fill: new ol.style.Fill({
                    color: 'rgba(255, 255, 0, 0.1)'
                })
            }),
            'Polygon': new ol.style.Style({
                stroke: new ol.style.Stroke({
                    color: 'blue',
                    lineDash: [4],
                    width: 3
                }),
                fill: new ol.style.Fill({
                    color: 'rgba(0, 0, 255, 0.1)'
                })
            }),
            'GeometryCollection': new ol.style.Style({
                stroke: new ol.style.Stroke({
                    color: 'magenta',
                    width: 2
                }),
                fill: new ol.style.Fill({
                    color: 'magenta'
                }),
                image: new ol.style.Circle({
                    radius: 5,
                    fill: null,
                    stroke: new ol.style.Stroke({
                        color: 'magenta'
```

```
            })
        })
    }),
    'Circle': new ol.style.Style({
        stroke: new ol.style.Stroke({
            color: 'red',
            width: 2
        }),
        fill: new ol.style.Fill({
            color: 'rgba(255,0,0,0.2)'
        })
    })
};

var styleFunction = function(feature) {
    return styles[feature.getGeometry().getType()];
};

var geojsonObject = {
    'type': 'FeatureCollection',
    'crs': {
        'type': 'name',
        'properties': {
            'name': 'EPSG:3857'
        }
    },
    'features': [{
        'type': 'Feature',
        'geometry': {
            'type': 'Point',
            'coordinates': [0, 0]
        }
    }, {
        'type': 'Feature',
        'geometry': {
            'type': 'LineString',
            'coordinates': [[4e6, -2e6], [8e6, 2e6]]
        }
    }, {
```

```
            'type': 'Feature',
            'geometry': {
                'type': 'LineString',
                'coordinates': [[4e6, 2e6], [8e6, -2e6]]
            }
        }, {
            'type': 'Feature',
            'geometry': {
                'type': 'Polygon',
                'coordinates': [[[-5e6, -1e6], [-4e6, 1e6], [-3e6, -1e6]]]
            }
        }, {
            'type': 'Feature',
            'geometry': {
                'type': 'MultiLineString',
                'coordinates': [
                    [[-1e6, -7.5e5], [-1e6, 7.5e5]],
                    [[1e6, -7.5e5], [1e6, 7.5e5]],
                    [[-7.5e5, -1e6], [7.5e5, -1e6]],
                    [[-7.5e5, 1e6], [7.5e5, 1e6]]
                ]
            }
        }, {
            'type': 'Feature',
            'geometry': {
                'type': 'MultiPolygon',
                'coordinates': [
                    [[[-5e6, 6e6], [-5e6, 8e6], [-3e6, 8e6], [-3e6, 6e6]]],
                    [[[-2e6, 6e6], [-2e6, 8e6], [0, 8e6], [0, 6e6]]],
                    [[[1e6, 6e6], [1e6, 8e6], [3e6, 8e6], [3e6, 6e6]]]
                ]
            }
        }, {
            'type': 'Feature',
            'geometry': {
                'type': 'GeometryCollection',
                'geometries': [{
                    'type': 'LineString',
                    'coordinates': [[-5e6, -5e6], [0, -5e6]]
```

```
            }, {
                'type': 'Point',
                'coordinates': [4e6, -5e6]
            }, {
                'type': 'Polygon',
                'coordinates': [[[1e6, -6e6], [2e6, -4e6], [3e6, -6e6]]]
            }]
        }
    }]
};

var vectorSource = new ol.source.Vector({
    features: (new ol.format.GeoJSON()).readFeatures(geojsonObject)
});

vectorSource.addFeature(new ol.Feature(new ol.geom.Circle([5e6, 7e6], 1e6)));

var vectorLayer = new ol.layer.Vector({
    source: vectorSource,
    style: styleFunction
});

var map = new ol.Map({
    layers: [
        new ol.layer.Tile({
            source: new ol.source.OSM()
        }),
        vectorLayer
    ],
    target: 'map',
    controls: ol.control.defaults({
        attributionOptions: ({
            collapsible: false
        })
    }),
    view: new ol.View({
        center: [0, 0],
        zoom: 2
    })
```

          });
        </script>
      </body>
</html>

实验 3-5 屏幕交互操作

本实验练习 OpenLayers 中封装的 Select、Draw、Modify 功能，实现空间查询、屏幕交互绘制点、线、面元素等编辑操作。

（1）实验目的：通过实验，认识 Select、Draw、Modify 的事件和操作流程，实现地图上要素的空间选择查询与编辑。

（2）相关实验：实验 3-4 地图图层叠加显示。

（3）实验数据：OSM 在线数据资源、GeoServer 示例数据。

（4）实验环境：GeoServer、Java JDK 、HBuilderX。

（5）实验内容：空间选择查询；交互绘制点、线、面，实现图层要素的动态添加。

1. 交互绘制

通过鼠标点击事件，实时交互绘制点、线、面等元素。利用 Draw 绘制，指定绘制的元素类型，然后设置所要添加的图层，将绘制的元素添加到图层中。

下面以绘制线元素为例，将新绘制的线添加到 lineLayer 图层中，示例代码如下：

```
.addInteraction(new ol.interaction.Draw({
        type: 'LineString',
        source: lineLayer.getSource()        // 绘制的线会添加到 source 中
    }));
```

自主练习：通过鼠标点击事件，交互绘制点、线等元素，结果对照图如图 3-5-1 所示。

在地图上通过鼠标点击交互绘制线的示例代码如下：

```
<!Doctype html>
<html xmlns=http://www.w3.org/1999/xhtml>
<head>
    <meta http-equiv=Content-Type content="text/html;charset=utf-8">
    <meta http-equiv=X-UA-Compatible content="IE=edge,chrome=1">
    <meta content=always name=referrer>
    <title>在地图上手动绘制线</title>
    <link rel="stylesheet" href="http://localhost:8090/geoserver/openlayers3/ol.css" type=" text/css">
    <script src="http://localhost:8090/geoserver/openlayers3/ol.js" type="text/javascript">
    </script>
</head>


<body>
```

```html
<div id="map" style="width: 100%"></div>
<script type="text/javascript">
    var map = new ol.Map({
        layers: [
            new ol.layer.Tile({
                source: new ol.source.OSM()
            })
        ],
        target: 'map',
        view: new ol.View({
            center: ol.proj.transform(
                [104, 30], 'EPSG:4326', 'EPSG:3857'),
            zoom: 10
        })
    });

    // 添加一个绘制的线的图层 lineLayer
    var lineLayer = new ol.layer.Vector({
        source: new ol.source.Vector(),
        style: new ol.style.Style({
            stroke: new ol.style.Stroke({
                color: 'green',
                size: 10
            })
        })
    })
    map2.addLayer(lineLayer);
    map2.addInteraction(new ol.interaction.Draw({
        type: 'LineString',
        source: lineLayer.getSource()        // 绘制的线会添加到 source 中
    }));
</script>
</body>
</html>
```

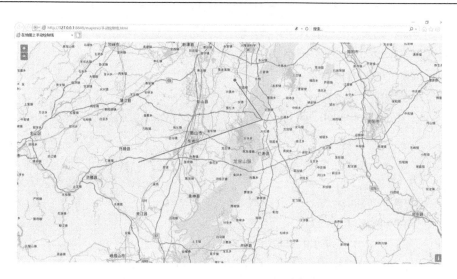

图 3-5-1　鼠标点击绘制线示例代码测试结果

2. 空间选择查询与编辑

首先发布 WFS 地图服务，然后将其作为数据源加载到程序中。将该数据作为矢量图形要素图层，实现空间选择与编辑。

数据加载时，service 类型应为 WFS，示例代码如下：

```
var wfsVectorLayer = new ol.layer.Vector({
    source: new ol.source.Vector({
        format: new ol.format.GeoJSON({
            geometryName: 'the_geom',
        }),
        url:
'http://localhost:8090/geoserver/nyc/ows?service=WFS&version=1.0.0&request=GetFeature&typeName=nyc%3Anyc_roads&outputFormat=application%2Fjson'
    }),
});
```

空间选择（Select）、图形编辑（Modify），在 OpenLayers 中已在交互功能的基类 interaction 中进行了实例化封装，可直接引入使用，也可以根据实际需求实现功能扩展。

空间选择（Select）返回的是鼠标点击获取的一个图形要素（Feature），针对选中的图形要素，可以通过鼠标拖动修改选中图形的节点坐标，实现图形编辑（Modify）。

调用交互操作的代码如下：

```
var select = new ol.interaction.Select();
var modify = new ol.interaction.Modify({
    features:select.getFeatures()
});
```

修改后，需要调用 writeTransaction 将数据传回服务器。

```
var WFSTSerializer = new ol.format.WFS();
var featObject = WFSTSerializer.writeTransaction(...)
```

参考示例代码如下：

```html
<!DOCTYPE html>
<html lang="en">
    <head>
        <meta http-equiv=Content-Type content="text/html;charset=utf-8">
        <meta http-equiv=X-UA-Compatible content="IE=edge,chrome=1">
        <meta content=always name=referrer>
        <title>WFS 图层</title>
        <link rel="stylesheet" href="http://localhost:8090/geoserver/openlayers3/ol.css"
          type="text/css">
        <script src="http://localhost:8090/geoserver/openlayers3/ol.js" type="
          text/javascript"></script>
    </head>
    <body>
        <input id="select" type="checkbox" value="select" checked />选择
        <input id="modify" type="checkbox" value="modify" />编辑
        <input id="name_input" type="input" />
        <input id="save" type="button" value="保存" onclick="onSave();" />
        <div id="map" style="height: 70%;">
        </div>

    </body>

    <script>
        var modifiedFeatures = null;
        var selectedFeature = null;
        var wfsVectorLayer = null;

        // 创建新的 WFS 图层
        var wfsVectorLayer = new ol.layer.Vector({
            source: new ol.source.Vector({
                format: new ol.format.GeoJSON({
                    geometryName: 'the_geom',
                }),
```

```
            url:
            'http://localhost:8090/geoserver/nyc/ows?service=WFS&version=1.0.0&
            request=GetFeature&typeName=nyc%3Anyc_roads&outputFormat=
            application%2Fjson'
        }),
        style: function(feature, resolution) {
            return new ol.style.Style({
                stroke: new ol.style.Stroke({
                    color: 'blue',
                    width: 5
                }),
                fill: new ol.style.Fill({ //图层符号颜色，以及透明度
                    color: 'rgba(0, 255, 0, 0.6)'
                })
            });
        }
    });

    var projection = new ol.proj.Projection({
        code: 'EPSG:2908',
        units: 'm',
    });
    var map = new ol.Map({
        view: new ol.View({
            projection: projection,
            center: [985079.32566908, 209690.38112203],
            maxZoom: 25,
            zoom: 18,
        }),
        target: 'map',
        layers: [wfsVectorLayer]
    })

    var style = new ol.style.Style({
        stroke: new ol.style.Stroke({
            color: 'red',
            width: 5
        }),
        fill: new ol.style.Fill({ //矢量图层符号颜色，以及透明度
```

```js
        color: 'rgba(0, 255, 0, 0.6)'
    })
});

var selectInteraction = new ol.interaction.Select({
    style: style
});

selectInteraction.on("select", function(e) {
    var features = e.selected;
    var feature = features[0];
    selectedFeature = feature;
    var attribute = feature.getProperties();
    var name = attribute["name"];

    document.getElementById("name_input").value = name;
});
map.addInteraction(selectInteraction);

var modifyInteraction = new ol.interaction.Modify({
    style: style,
    features: selectInteraction.getFeatures()
});

modifyInteraction.on('modifyend', function(p1) {
    modifiedFeatures = p1.features;
})

document.getElementById('select').onchange = function() {
    if (this.checked) {
        // 勾选选择复选框时，添加选择器到地图
        map.removeInteraction(selectInteraction);
        map.addInteraction(selectInteraction);
    } else {
        // 不勾选选择复选框的情况下，移出选择器和修改器
        map.removeInteraction(selectInteraction);
        document.getElementById('modify').checked = false;
        map.removeInteraction(modifyInteraction);
        modifiedFeatures = null;
```

```javascript
        }
    };

    document.getElementById('modify').onchange = function() {
        if (this.checked) {
            // 勾选修改复选框时，添加选择器和修改器到地图
            document.getElementById('select').checked = true;
            map.removeInteraction(modifyInteraction);
            map.addInteraction(modifyInteraction);
            map.removeInteraction(selectInteraction);
            map.addInteraction(selectInteraction);
        } else {
            // 不勾选修改复选框时，移出修改器
            map.removeInteraction(modifyInteraction);
            modifiedFeatures = null;
        }
    };

    // 保存已经编辑的要素
    function onSave() {
        if (selectedFeature != null) {
            var newName = document.getElementById("name_input").value;
            selectedFeature.set("name", newName);
            modifyWfs([selectedFeature]);
        }
        if (modifiedFeatures && modifiedFeatures.getLength() > 0) {
            var modifiedFeature = modifiedFeatures.item(0).clone();
            // 注意 ID 是必须的，通过 ID 才能找到对应修改的 feature
            modifiedFeature.setId(modifiedFeatures.item(0).getId());
            modifyWfs([modifiedFeature]);
        }
    }

    // 将修改后的数据提交到服务器端
    function modifyWfs(features) {
        var WFSTSerializer = new ol.format.WFS();
        var featObject = WFSTSerializer.writeTransaction(null,
            features, null, {
                featureType: 'nyc_roads', //图层名
```

// 注意这个值必须是创建工作区时的命名空间 URL
                    featureNS: 'http://geoserver.org/nyc',
                    srsName: 'EPSG:2908'
            });
            // 转换为 xml 内容发送到服务器端
            var serializer = new XMLSerializer();
            var featString = serializer.serializeToString(featObject);
            var request = new XMLHttpRequest();
            request.open('POST', 'http://localhost:8090/geoserver/nyc/ows?service=WFS');
            // 指定内容为 xml 类型
            request.setRequestHeader('Content-Type', 'text/xml');
            request.send(featString);
        }
    </script>
</html>

运行该代码，图形编辑效果如图 3-5-2 所示。

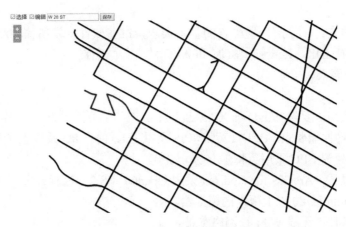

图 3-5-2　空间选择与编辑界面

自主练习：编码实现通过鼠标点击选中地图要素，并通过鼠标拖动编辑图形的节点，编辑完成后上传保存数据。

第四部分 基于 ArcGIS 的 Android 移动 GIS 开发系列实验

实验 4-1 ArcGIS SDK for Android 开发环境及配置

ArcGIS SDK for Android 开发环境的实质是"Android 开发环境"+"ArcGIS 开发控件"，而 Android 开发环境是基于 Eclipse 的。因此，本实验的主要目的是了解 ArcGIS SDK for Android 开发环境搭建的步骤。

（1）实验目的：准备相关的软件，了解环境配置的基本要求，并逐步完成开发环境的搭建，为接下来的实习做好准备。

（2）相关实验：GIS 专业实验设备与环境配置中的"GIS 应用开发环境"和"GIS 应用开发资源"。

（3）实验数据：本教材系列实验数据。

（4）实验环境：Eclipse 3.6.2（Helios）以上、JDK6 及以上、ArcGIS for Server、Android 2.2 及以上。

（5）实验内容：①了解环境配置的基本要求，准备软件开发所需的相关软件；②完成环境的搭建，包括 JDK、Eclipse、SDK 和 AVD 的安装与配置。

1. 实习环境要求

1）系统环境要求

（1）操作系统：要求 Windows XP 系统以上。

（2）PC 硬件性能要求：CPU（英特尔 i5 及以上）、内存 8G 及以上（若使用 Android 虚拟机进行开发，建议 CPU 采用英特尔 i5 及以上）。

（3）移动设备：ArcGIS for Android 采用 OpenGL ES 2.0 进行开发。如果使用模拟器开发，建议采用 Android 4.0 以上版本的模拟器。

（4）网络条件：无线 WiFi + 网络宽带。

（5）ArcGIS for Server：版本 10.1。

2）开发环境要求

（1）Eclipse 3.6.2（Helios）以上（本实验版本为 4.2.0）。

（2）Eclipse JDT 插件（已在大多数 Eclipse 包中安装）。

（3）Eclipse 有用于多种开发目的的软件包，建议使用的软件包为 Eclipse IDE for Java EE Developers；Eclipse IDE for Java Developers；Eclipse Classic。

（4）JDK6 及以上版本。

（5）在安装 ESRI 提供的开发插件之前，需要安装 ADT 插件。

3）Android 系统要求

SDK 平台要求 Android 2.2 及以上，API 8 及以上。

2. JDK 的安装配置

通过相关网络途径下载 JDK1.8.0，具体安装过程如下。

（1）下载安装包后解压。

（2）点击安装，根据安装提示完成，安装完毕后点击"关闭"。

（3）设置系统环境变量：右击"我的电脑"→"属性"→"高级系统设置"→"环境变量"，如图 4-1-1 所示。

图 4-1-1　系统环境变量设置界面

之后进入环境变量设置窗口，如图 4-1-2 所示。在系统变量中点击"新建"，设置新建的系统变量名为"JAVA_HOME"，变量值是 JDK 的安装路径下的 bin 文件夹（本例为 C:\Program Files\Java\jdk1.8.0_65\bin），如图 4-1-3 所示。

图 4-1-2　环境变量设置窗口

图 4-1-3 新建系统变量示例图

在"系统变量"中寻找"Path"变量并点击"编辑",在变量值最后输入"%JAVA_HOME%\bin;%JAVA_HOME%\jre\bin;"。

注意:如果原"Path"的变量值末尾没有";"号,先添加";"号,然后再输入上面的变量值,如图 4-1-4 所示。

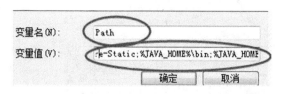

图 4-1-4 Path 变量编辑界面

在"系统变量"中继续新建一个"CLASSPATH"变量,变量值填写".;%JAVA_HOME%\lib;%JAVA_HOME%\lib\tools.jar"(注意最前面有个"."),如图 4-1-5 所示。

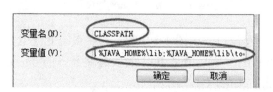

图 4-1-5 CLASSPATH 变量编辑界面

点击"确定"后系统变量设置完毕。

运行 cmd 输入 java -version(注意:java 和-version 之间有一个空格),检验是否配置成功。若如图 4-1-6 所示显示版本信息则说明安装和配置成功。

图 4-1-6 java 版本检验界面

3. Eclipse、Android 开发环境(ADT)的安装配置

要求 Eclipse 版本为 3.6.2 以上,推荐 3.7 版本或者 4.2 版本(本实验版本为 4.2)。本实验

中以下载的 ADT 集成软件（也就是已经安装好 ADT 插件的 Eclipse）"adt-bundle-windows-x86-20131030"搭建 Android 开发环境。

将该压缩包直接解压到硬盘里便可使用。解压后的文件内容包括 Eclipse 文件、sdk 文件和 SDK 管理应用程序。

新建一个 workspace 文件夹，该文件夹用以存储实习中创建的 Android 项目工程，如图 4-1-7 所示。

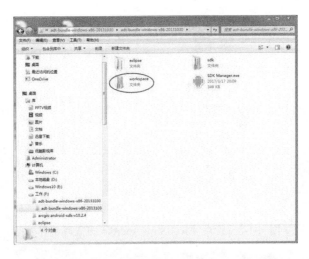

图 4-1-7　ADT 集成软件与 workspace 工作路径

打开"eclipse"文件夹，运行 eclipse.exe。自此"adt-bundle-windows-x86"安装完成。

下面设置"工作路径"。打开"File"→"switch workspace"→"other"，点击"Browse"选择上述步骤中新建的"workspace"文件夹作为工作路径，点击"确定"，点击"OK"，等待界面重启。

至此 Android 开发环境就已经建立好，接下来可以新建项目了。

4. ArcGIS Runtime SDK for Android 的版本要求及安装配置

从 ArcGIS for Android SDK 10.2.5 版本开始，ESRI 不再提供 Eclipse 的插件支持，官方的帮助也只针对 Android Studio 的支持。在此，本实验提供 10.2.4 版本 ArcGIS Android SDK 的环境配置。ArcGIS Runtime SDK for Android 的安装可以分为在线安装和离线安装两种方式。

1）在线安装

通常最简便的方式是在 eclipse 中进行在线安装，ArcGIS Developer 网站提供 ArcGIS Runtime SDK for Android 的在线安装，地址为：https://developers.arcgis.com/android/。

安装过程：打开 eclipse，点击"Help"→"Install new software"→"Add"，然后在"work with"中输入地址，在"name"中输入"ArcGIS"，点击"ok"，选中 ArcGIS Android 插件压缩包，点击"Next"，依步骤完成安装即可。

2）离线安装

本实验以 arcgis-android-sdk-v10.2.4 版本为例。首先下载离线包，然后进行解压安装。

打开 eclipse，点击"Help"→"Install new software"→"Add"，然后在"name"中输入

"ArcGIS",点击"Archive"按钮,找到 jar 包安装路径(例如,…\tools\eclipse-plugin\arcgis-android-eclipse-plugin.jar),点击"ok",选中 ArcGIS Android 插件压缩包,点击"Next",依步骤完成安装即可。

安装完成并重启 Eclipse 后,打开菜单"File"→"New"→"Project…",可以看到在 New Project 中,已经有"ArcGIS for Android"可供选择,说明 ArcGIS for Android 已经安装成功,如图 4-1-8 所示。

图 4-1-8 检查 ArcGIS for Android 开发环境界面

5. Android 模拟器(AVD)配置

打开 Eclipse,选择"Windows"→"Android Virtual Device Manager"→"new",打开"Create new Android Virtual Device(AVD)"对话框,如图 4-1-9 所示,在该对话框中配置"AVD"的名称、CPU、内存等相关属性。点击"OK",Android 模拟器即创建完成。

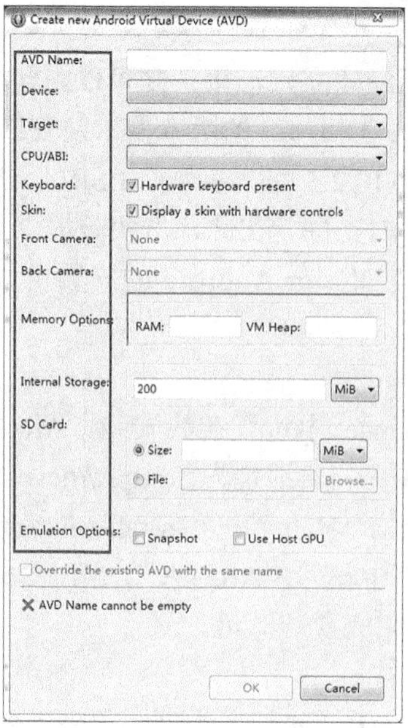

图 4-1-9 创建新 AVD 窗口

实验 4-2 地图工程创建

地图显示是移动 GIS 的最基本功能。通过本节实验，了解开发环境，熟悉 Eclipse 的结构，实现地图数据的加载和显示。

（1）实验目的：通过实验 4-1，我们已经了解了如何配置 ArcGIS for Android 开发环境。本实验通过使用 Eclipse，创建一个 ArcGIS 移动项目，了解 Eclipse 的结构、组成及相关功能。

（2）相关实验：实验 4-1 ArcGIS SDK for Android 开发环境及配置。

（3）实验数据：本教材系列实验数据。

（4）实验环境：Eclipse4.2、ArcGIS for Android 插件、JAVA 编程语言。

（5）实验内容：通过使用 Eclipse，创建一个 ArcGIS 移动项目"Hello World Map"，也就是创建一个地图工程，实现地图数据的显示功能，并且分析项目的相关结构。

1. 创建"Hello World Map"工程

创建工程项目的具体步骤如下。

（1）打开我们已经配置好的 Eclipse 工具，点击"File"→"New…"→"Other…"，在弹出的窗体中找到并选中 ArcGIS for Android 下的"ArcGIS Project for Android"，然后点击"Next"，在弹出的对话框中工程名称"Project Name"输入框中输入要创建的项目名"Hello WorldMap"，然后点击"Next"，之后跳转到应用设置界面，修改"Package Name"输入框的包结构名，如：Esri.GIS.Demo，修改完成后，点击"Finish"。其步骤如图 4-2-1 所示。

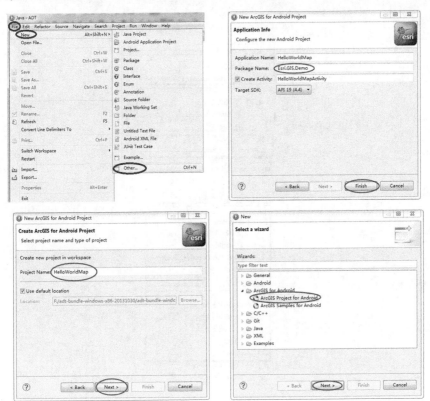

图 4-2-1　创建新工程步骤图

（2）点击"Finish"之后 Eclipse 将自动完成 ArcGIS 开发包的引用、程序基本代码等工程的组织工作。此时我们新创建了一个项目，其工程结构如图 4-2-2 所示。

图 4-2-2　新建项目工程结构示意图

2. 地图项目工程结构

根据图 4-2-2，可以发现 ArcGIS 项目与普通的 Android 项目基本相同。下面整体介绍 ArcGIS 的项目工程结构，其分类图如图 4-2-3 所示。

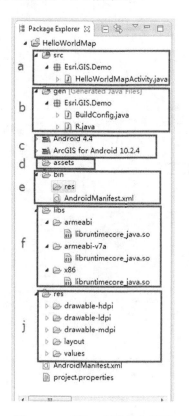

图 4-2-3　项目工程结构分类图

图中 a 区域是存放 java 源码的目录，目录中的文件是根据 package 结构管理的。

图中 b 区域里的 gen 也是一个源代码目录，但其中的 Java 文件是由 Android 平台自动生成的，这个目录下的 R.java 类文件是由 Android Framework 负责管理的，不需用户手动操作。在 gen 中有一个 BuildConfig.java 文件，它的作用是检查用户所编写的代码并运行调试。

图中 c 区域是项目中所需要的 java 函数库，这比普通的 Android 项目多了 ArcGIS 的函数库。

图中 d 区域表示可以将所需的文件放在 assets 目录中进行存储，方便访问。assets 中的资源文件与 res 中的功能很相似，都是存放资源文件的目录，但 assets 中的资源不会像 res 中的资源那样为每个资源文件生成 ID 标识。

图中 e 区域目录是存放编译后生成的应用程序。

图中 f 区域的 libs 目录下存放的是一些项目所需的动态链接库。对于 ArcGIS 项目它默认存放了两个 GIS 所需的动态链接库，当然也可以添加一些用户所需的其他动态链接库。

图中 j 区域 res 目录存放了用户所需的大部分资源。默认目录下有三类资源：drawable 目录主要存放一些图片、layout 目录主要放一些布局文件、values 目录主要存放一些项目中所需的参数值文件；当然除了这些还有其他分类，如 anim 和 xml 目录等，在此不再一一介绍。

最后的 AndroidManifest.xml 文件是项目的一个系统配置文件，它包含了 activity（行为）、view（视图）、service（服务）之类的信息，以及运行这个 Android 应用程序需要的用户权限列表，同时也详细描述了 Android 应用的项目结构。

3. 地图工程的编码实现

介绍完 ArcGIS 的项目结构，接下来通过代码实例演示地图工程的运行，了解如何才能正常显示地图。

打开"HelloWorldMapActivity.java"文件，修改代码如下：

```
public void onCreate(Bundle savedInstanceState) {
    super.onCreate(savedInstanceState);
    mMapView = new MapView(this);//实例化 MapView 对象
    mMapView.setLayoutParams(new LayoutParams(LayoutParams.FILL_PARENT, LayoutParams.FILL_PARENT));
    ArcGISTiledMapServiceLayer tileLayer = new ArcGISTiledMapServiceLayer("https://services.arcgisonline.com/arcgis/rest/services/ESRI_Imagery_World_2D/MapServer");//实例化图层
    mMapView.addLayer(tileLayer);//添加图层
    SetContentView(mMapView);
}
```

修改保存后，按照错误提示器修改代码中的错误，检查无误后连接 Android 手机设备或者运行 AVD，然后右击"Run as"→"Android Application"，选择可运行的设备，运行结果如图 4-2-4 所示。

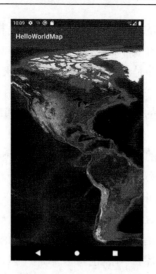

图 4-2-4　示例运行结果界面

自主练习：创建一个新的地图工程，添加地图数据并模拟预览。

实验 4-3　数据显示与浏览

MapView 控件是 ArcGIS Runtime SDK for Android 的核心组件。通过 MapView 可以呈现地图数据。在 MapView 中定义了丰富的属性、方法和事件，用户通过 MapView 可以操作设备的触摸屏。默认 MapView 可以响应用户的手势操作。在 GIS 的开发中，常见的地图操作有缩放，旋转，平移，获取范围、比例尺、分辨率等信息。

（1）实验目的：掌握 MapView 控件的功能和特性，实现地图常见的操作；理解地图图层加载的过程；熟悉监听器的使用方法；掌握空间要素可视化，实现对地图视图的基本浏览操作。

（2）相关实验：实验 4-2 创建地图工程。

（3）实验数据：本教材系列实验数据。

（4）实验环境：Eclipse4.2、ArcGIS for Android 插件、JAVA 编程语言。

（5）实验内容：通过 MapView 控件实现地图常见的操作，例如缩放，旋转，平移，获取范围、比例尺、分辨率等信息。

1. MapView 控件初识

MapView 具有呈现地图的能力。MapView 中可以添加一个或多个图层，支持多种类型。地图数据只有作为图层添加到 MapView 容器中才能显示。

通过 MapView 可以设置地图的显示范围、是否允许地图旋转、设置地图背景、地图的最大/最小分辨率等。

MapView 提供的丰富的手势监听接口，通过这些监听器，可以监听各种手势动作，如点击、双击、移动或长按等操作。

2. MapView 控件的添加方式

MapView 控件的添加方式有 XML 和硬编码两种。一般多采用 XML 方式，方便调整布局及其属性相关设置。

1）XML 方式

XML 方式的代码如下：

```xml
<com.esri.android.map.MapView
        android:id="@+id/map"
        android:layout_width="fill_parent"
        android:layout_height="match_parent">
</com.esri.android.map.MapView>
```

2）硬编码方式

直接在 Java 文件中加载 MapView，载入地图服务，代码如下：

```java
MapView map = new MapView(this);
        map.setLayoutParams(new LayoutParams(LayoutParams.FILL_PARENT, LayoutParams.FILL_PARENT));
        tileLayer = new ArcGISTiledMapServiceLayer("http://services.arcgisonline.com/ArcGIS/rest/services/World_Street_Map/MapServer");
        map.addLayer(tileLayer);
        setContentView(map);
```

3. MapView 控件的功能

ArcGIS for Android 插件中，MapView 直接继承自 Android 的 ViewGroup。因此 MapView 类继承了 ViewGroup 的所有方法和属性，其操作方式同 ViewGroup 极其相似。

注意：在下面的操作练习中，需要事先将地图加载到 MapView。

1）地图显示基本设置

在 ArcGIS for Android 中，MapView 具有很多与地图显示设置有关的方法，其中与地图显示的比例尺、分辨率、中心点、范围有关的方法如表 4-3-1 所示。

表 4-3-1 地图显示基本设置方法

返回类型	方法	说明
Void	centerAt(Point centerPt, Boolean animated)	以指定的点为中心，将地图居中显示
Point	getCenter()	获取地图中心点
Polygon	getExtent()	获取地图范围的最小外包矩形
Envelope	getMapBoundaryExtent()	获取地图的边界
Void	setExtent(Geometry geometry)	将地图放大到指定的范围，将该指定的几何图形（geometry）的边界（bound）作为地图当前的显示范围（extent）
Double	getMaxResolution()	获取地图最大分辨率
Void	setMaxResolution(double maxResolution)	设置地图最大分辨率
Double	getMinResolution()	获取地图最小分辨率
Void	setMinResolution(double minResolution)	设置地图最小分辨率
Double	getResolution()	获取当前地图分辨率

续表

返回类型	方法	说明
Void	setResolution(double res)	设置当前地图分辨率
Double	getScale()	获取当前地图比例尺
Void	setScale(double scale)	设置当前地图比例尺

要获取/设置地图的比例尺、初始分辨率、范围、中心点等信息，直接使用上述方法即可。部分控件的使用方法如下：

map.getCenter();//获取地图中心点；

map.getScale();//获取当前地图比例尺；

map.setScale(18000000);//设置地图初始化时的比例尺分母；

map.setExtent(new Envelope(371987, 252920, 624459, 423400));//范围；

2）地图缩放、平移和旋转

与缩放、平移和旋转有关的地图事件如表 4-3-2 所示。

表 4-3-2　地图缩放、平移和旋转方法

返回类型	方法	说明
Void	zoomin()	地图放大
Void	zoomout()	地图缩小
Void	zoomTo(Point centerPt, float factor)	将地图放大到指定点
Void	zoomToResolution(Point centerPt, double res)	将地图放大到指定分辨率
Void	zoomToScale(Point centerPt, double scale)	将地图放大到指定比例尺
Double	getRotationAngle()	获取当前地图旋转角度（单位为°）
Void	setRotationAngle(double degree)	将地图按照指定的角度（单位为°）旋转，逆时针方向角度度数为正
Void	setRotationAngle(double degree, float pivotX, float pivotY)	将地图按指定的点和角度旋转，逆时针方向角度度数为正
Void	setAllowRotationByPinch(boolean allowRotationByPinch)	设置允许/取消手势旋转，即根据两个手指头的旋转来使地图跟着旋转
Boolean	isAllowRotationByPinch()	获取是否允许手势旋转

（1）平移。MapView 中已经默认支持平移操作，即使用鼠标或手势拖动地图时就会平移地图。

（2）缩放。在 ArcGIS for Android 中，通过分辨率或比例尺来控制缩放级别。可以用 getResolution()和 getscale()方法获取当前地图的比例尺和分辨率，然后使用 zoomTo()/ zoomToScale()/zoomToResolution()达到控制地图缩放级别的目的。

示例代码如下：

　　map.zoomin();

　　map.zoomout();

　　map.zoomTo(point centerPt, float factor);

map.zoomToResolution(point centerPt, double res);

map.zoomToScale(Point centerPt, double scale);

其中,前两种功能是逐级缩放,即表示调用一次方法,地图将放大或者缩小一级。在 zoomTo(point centerPt, float factor)中,centerPt 指以指定点为显示中心进行放大,factor 参数用以计算新的分辨率,计算公式为:新的分辨率 = 当前分辨率/factor(例如,如果期望将地图在当前分辨率下放大 3 级,则新分辨率 = 当前分辨率/8,因为每一级之间分辨率是 2 的倍数关系,放大 3 级,factor =2^3)。

在 zoomToScale(Point centerPt, double scale)和 zoomToResolution(point centerPt, double res)中,scale 和 res 分别指实际的分辨率和比例尺,按照缩放级别通过 2 的倍数关系直接计算即可。

(3)最大最小缩放级别。设置地图最大最小缩放级别的示例代码如下:

map.setMaxResolution(MaxResolution);

map.setMinResolution(MinResolution);

上述代码设置了地图的最大、最小分辨率,也就限制了地图的缩放级别。当地图达到最大分辨率时,地图将不能再放大;当地图达到最小分辨率时,地图将不能再缩小。

(4)旋转地图。可以使用 setRotationAngle(double degree)和 setRotationAngle (double degree, float pivotX, float pivotY)实现地图旋转。如果要实现手势旋转,需要通过 setOnPinchListener (OnPinchListeneronPinchListener)监听来实现。

示例代码如下:

map.setAllowRotationByPinch(true); //是否允许使用手势方式旋转地图

map.setRotationAngle(15.0); //初始化时将地图旋转 15°,逆时针方向为正

3)获取地图上某点的坐标

获取地图上某点的坐标使用下列几种方法,其中,主要使用 toMapPoint()方法实现获取地图上的点坐标信息,具体见表 4-3-3。

表 4-3-3 获取地图上点的坐标方法

返回类型	方法	说明
SpatialReference	getSpatialReference()	获取当前地图的坐标系统
Point	toMapPoint(float screenx, float screeny)	将屏幕上某点的坐标转换成地图坐标系下的点坐标(ArcGIS geometry Point)
Point	toMapPoint(Point src)	将屏幕上某点的坐标转换成地图坐标系下的点坐标(ArcGIS geometry Point)
Point	toScreenPoint(Point src)	将地图坐标系下的点坐标(ArcGIS geometry Point)转换成屏幕坐标

4)监听手势操作

移动端手势的监听是一个重要的环节,地图的手势操作由 MapView 来管理,主要响应手势,如表 4-3-4 所示。

表 4-3-4 手势监听操作

返回类型	方法/事件监听	说明
OnLongPressListener	getOnLongPressListener()	获取地图长按事件监听
OnPanListener	getOnPanListener()	获取地图平移事件监听
OnPinchListener	getOnPinchListener()	获取地图捏夹事件监听
OnSingleTapListener	getOnSingleTapListener()	获取地图点击事件监听
OnZoomListener	getOnZoomListener()	获取缩放监听
Void	setOnLongPressListener(OnLongPressListeneronLongPressListener)	设置地图长按事件监听
Void	setOnPanListener(OnPanListeneronPanListener)	设置地图平移事件监听
Void	setOnPinchListener(OnPinchListeneronPinchListener)	设置地图捏夹事件监听
Void	setOnSingleTapListener(OnSingleTapListeneronSingleTapListener)	设置地图点击事件监听
Void	setOnZoomListener(OnZoomListeneronZoomListener)	设置缩放监听

利用这些监听，可以实现用户对地图的多种交互操作。

4. 加载 Layer 地图图层

图层（Layer）是空间数据的载体。在 Android 移动 GIS 的开发中，只有将图层添加到 MapView 对象中，才能浏览显示所加载的空间数据。图层类型有多种，如图 4-3-1 所示。

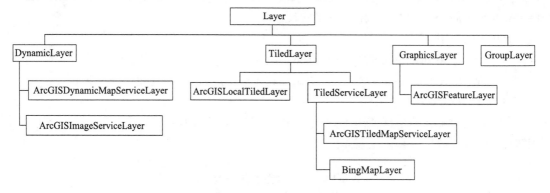

图 4-3-1 图层结构关系图

1）ArcGISTiledMapServiceLayer

在 ArcGIS Server 中可以发布多种地图服务。开发过程中，要根据地图服务的种类，选择相应类型的图层来获取应用这些服务。

ArcGISTiledMapServiceLayer 图层，对应 ArcGIS Server 服务中的"切片数据服务"。在程序开发中，ArcGISTiledMapServiceLayer 的调用方法如下：

MapView mv = new MapView(this);

mv.addLayer(new ArcGISTiledMapServiceLayer("http://services.arcgisonline.com/ArcGIS/rest/services/World_Topo_Map/MapServer"));

setContentView(mv);

2）ArcGISDynamicMapServiceLayer

ArcGISDynamicMapServiceLayer 图层，对应 ArcGIS Server 服务中的动态服务，属于 DynamicLayer 的子类。ArcGISDynamicMapServiceLayer 的调用方法如下：

MapView mv = new MapView(this);

mv.addLayer(new ArcGISDynamicMapServiceLayer("http://sampleserver1.arcgisonline.com/ArcGIS/rest/services/Demographics/ESRI_Population_World/MapServer"));

setContentView(mv);

3）ArcGISImageServiceLayer

ArcGISImageServiceLayer 图层，对应 ArcGIS Server 服务中的影像数据服务，在程序开发中，调用影像数据服务的用法如下：

MapView mv = new MapView(this);

mv.addLayer(new ArcGISImageServiceLayer(

"http://myserver/arcgis/rest/services/MyImage/ImageServer",null));

setContentView(mv);

4）ArcGISFeatureLayer

ArcGISFeatureLayer 图层对应 ArcGIS Server 服务中的矢量数据服务（Feature Service），与其他图层类型相比，该类型具有更为丰富的功能，它继承自 GraphicsLayer，可以进行在线数据编辑，支持对图层的所有操作功能。

Feature Service 调用模式有三种，不同的模式返回的数据和响应效率不同。三种模式分别为：①MODE.SNAPSHOT，按快照方式返回数据；②MODE.ONDEMAND，按需返回数据；③MODE.SELECTION，按所选的范围返回数据；默认为 SNAPSHOT 模式。

ArcGISFeatureLayer 图层用法的示例代码如下：

Stringurl="https://servicesbeta.esri.com/ArcGIS/rest/services/SanJuan/TrailConditions/FeatureServer/0";

MapView mv= new MapView(this);

mv.addLayer(new ArcGISFeatureLayer(url,MODE.SNAPSHOT));//按照快照模式返回数据

setContentView(mv);

5）ArcGISLocalTiledLayer

ArcGISLocalTiledLayer 是添加离线数据包的图层，该图层目前支持缓存切片图层数据和 tpk 格式的数据。

以加载 tpk 格式的数据为例，基本步骤如下：

（1）打包切片地图为 tpk 格式。

（2）将 tpk 文件复制到终端设备中，记录下 tpk 在终端设备存储卡中的位置（建议新建一个文件夹，将 tpk 包放到该文件夹中）。

（3）在 addLayer 中进行设定。

示例代码如下：

MapView mv = new MapView(this);

ArcGISLocalTiledLayer   local = new

ArcGISLocalTiledLayer("file:///storage/sdcard1/arcgis/testmap.tpk");

mv.addLayer(local);

setContentView(mv);

如果添加缓存切片图层数据，示例代码如下：

ArcGISLocalTiledLayer local = new ArcGISLocalTiledLayer("file:/// mnt/ sdcard/ … / Layers");//离线缓存切片图层数据

6) GraphicsLayer

GraphicsLayer 可以包含一个或多个 Graphic 对象，所以无论是查询结果还是自己标绘的 Graphic 数据都要通过它来呈现。需要注意的是，GraphicsLayer 不能作为第一个图层添加到 MapView 中，因为 MapView 加载图层时首先要初始化一些地图参数，而 GraphicsLayer 图层不具备这些参数。

该类图层用法的示例代码如下：

MapView mv = new MapView(this);

mv.addLayer(new GraphicsLayer());

setContentView(mv);

除了可以呈现 Graphic 对象外，GraphicsLayer 还具备要素更新与要素获取等其他重要功能。

（1）要素更新。在移动设备上想要实现标绘时，就会用到 GraphicsLayer 的 updateGraphic() 方法进行地图实时更新。

示例代码如下：

```
public boolean onDragPointerMove(MotionEvent from, MotionEvent to) {
    if (startPoint == null) {
        graphicsLayer.removeAll();
        poly = new Polyline();
        startPoint = mapView.toMapPoint(from.getX(), from.getY());
        poly.startPath((float) startPoint.getX(),
            (float) startPoint.getY());
        uid = graphicsLayer.addGraphic(new Graphic(poly,new SimpleLineSymbol(Color.BLUE,5)));
    }
    poly.lineTo((float) mapPt.getX(), (float) mapPt.getY());
    graphicsLayer. updateGraphic(uid,poly);//更新显示
    return true;
}
```

（2）要素获取。ArcGIS Runtime for Android 与其他 Web API 有所不同，其他 API 中 Graphic 对象是可以设置监听的，而在 ArcGIS Runtime for Android 中的 Graphic 不能添加监听，所以当在地图上点击一个 Graphic 对象时就需要通过其他方式间接地获取这个对象。可以通过 GraphicsLayer 中的 getGraphicIDs(float x, float y, int tolerance)方法来获取要素，其中 x 和 y 是屏幕坐标，tolerance 是容差，通过这个方法就可以间接地获取所需的 Graphic 对象。

示例代码如下：
public boolean onSingleTap(MotionEvent e) {
　　　　Graphic graphic = **new** Graphic(mapView.toMapPoint(**new** Point(e.getX(),e.getY())),**new** SimpleMarkerSymbol(Color.RED,25,STYLE.CIRCLE));
　　　　return false;
　　　　int[] getGraphicIDs(**float** x,**float** y, **int** tolerance)
　　}

5. 监听器的使用

ArcGIS Runtime SDK for Android 为我们提供了丰富的事件监听器，本节主要介绍我们经常使用的监听器以及它们可以实现的功能，在如图 4-3-2 所示的监听器中只有 MapOnTouchListener 是类，其他皆为接口类型。

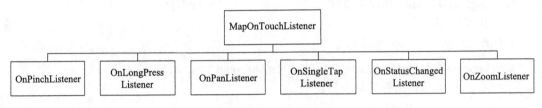

图 4-3-2　事件监听器

1) MapOnTouchListener

MapOnTouchListener 是 MapView 最为重要的监听器之一，它实现了 OnTouchListener 和 MapGestureDetector.OnGestureListener 接口。对于地图的所有操作 MapOnTouchListener 都可以进行响应，使用非常方便，在使用前只需扩展这个类并重写该类中的方法即可。

示例代码如下：
myListener = **new** MyTouchListener(**this**, mapView);
　　　mapView.setOnTouchListener(myListener);
　　class MyTouchListener **extends** MapOnTouchListener {
　　　　public MyTouchListener(Context context, MapView view) {
　　　　　　super(context, view);
　　　　}
　　　　public void setType(String geometryType) {
　　　　　　this.type = geometryType;
　　　　}
　　　　public String getType() {
　　　　　　return this.type;
　　　　}
　　　　public boolean onSingleTap(MotionEvent e) {
　　　　　　return true;
　　　　}
　　　　public boolean onDragPointerMove(MotionEvent from, MotionEvent to) {

```
            return super.onDragPointerMove(from, to);
        }
        @Override
        public boolean onDragPointerUp(MotionEvent from, MotionEvent to) {
            return super.onDragPointerUp(from, to);
        }
}
```
通过上面的代码就可以监听到手势操作，对于不同的手势操作将执行相对应的方法。例如，只需在 onSingleTap() 方法中完成点的获取、窗体的创建及其弹出操作即可实现在地图上点击即时弹出窗体的功能。

2）OnLongPressListener

OnLongPressListener 接口主要用于监听在地图上的长按事件。

示例用法如下：

```
//为地图添加一个长按监听器
mapView.setOnLongClickListener(new View.OnLongClickListener() {
        //长按后自动执行的方法
        public boolean onLongClick(View v) {
            return false;
        }
});
```

3）OnPanListener

OnPanListener 接口用于 MapView 平移地图操作时的事件监听。

示例用法如下：

```
//为地图添加一个平移监听器
    mapView.setOnPanListener(new OnPanListener() {
        public void prePointerUp(float fromx, float fromy, float tox, float toy) {
        }
        public void prePointerMove(float fromx, float fromy, float tox, float toy) {
        }
        public void postPointerUp(float fromx, float fromy, float tox, float toy) {
        }
        public void postPointerMove(float fromx, float fromy, float tox, float toy) {
        }
    });
```

4）OnPinchListener

OnPinchListener 接口是对地图进行两指或多指操作时用到的事件监听。

以通过该接口实现两指捏夹地图缩放为例，示例代码如下：

```
//添加捏夹监听器
    mapView.setOnPinchListener(new OnPinchListener() {
```

```java
            public void prePointersUp(float x1, float y1, float x2, float y2,
                    double factor) {
            }
            public void prePointersMove(float x1, float y1, float x2, float y2,
                    double factor) {
            }
            public void prePointersDown(float x1, float y1, float x2, float y2,
                    double factor) {
            }
            public void postPointersUp(float x1, float y1, float x2, float y2,
                    double factor) {
            }
            public void postPointersMove(float x1, float y1, float x2, float y2,
                    double factor) {
            }
            public void postPointersDown(float x1, float y1, float x2, float y2,
                    double factor) {
            }
        });
```

5）OnSingleTapListener

OnSingleTapListener 接口是对地图进行点击操作时的事件监听器。

示例用法如下：

```java
//为地图添加点击事件监听
        mapView.setOnSingleTapListener(new OnSingleTapListener() {
                //点击地图后自动执行的方法
                public void onSingleTap(float x, float y) {
                    // TODO Auto-generated method stub
                }
        });
```

6）OnStatusChangedListener

OnStatusChangedListener 接口用于监听 MapView 或 Layer 状态变化的监听器。

示例用法如下：

```java
//添加状态监听器
        mapView.setOnStatusChangedListener(new OnStatusChangedListener() {
                public void onStatusChanged(Object source, STATUS status) {
                    if(status == STATUS.INITIALIZED){
                    }else if(status == STATUS.LAYER_LOADED){
                    }else if((status == STATUS.INITIALIZATION_FAILED)){
                    }else if((status == STATUS.LAYER_LOADING_FAILED)){
```

                }
            }
        });

从上面的代码可以清晰看到，MapView 的状态变化主要有四种：

（1）STATUS.INITIALIZED 初始化成功。

（2）STATUS.LAYER_LOADED 图层加载成功。

（3）STATUS.INITIALIZATION_FAILED 初始化失败。

（4）STATUS.LAYER_LOADING_FAILED 图层加载失败。

7）OnZoomListener

OnZoomListener 接口主要监听地图的缩放事件。

示例用法如下：

```
mapView.setOnZoomListener(new OnZoomListener() {
        //缩放之前自动调用的方法
        public void preAction(float pivotX, float pivotY, double factor) {
        }
        //缩放之后自动调用的方法
        public void postAction(float pivotX, float pivotY, double factor) {
        }
});
```

6. 空间要素可视化

在 ArcGIS Runtime for Android 中，一般将空间要素添加到 GraphicsLayer 图层中进行展示。本节主要介绍 Graphic 的两个组成部分：Geomtry 和 Symbol，以及 Renderer 接口。

1）Graphic

Graphic 是承载空间几何要素的载体，Graphic 对象可以添加到 GraphicsLayer 图层中进行展示。Graphic 主要由四部分组成：Geomtry、symbol、Map 和 InfoTemplate。通过 Graphic 对象可以获取这四部分相应的属性。如果要修改属性，可以通过使用 GraphicsLayer 的 updateGraphic()方法间接实现。

示例用法如下：

```
public boolean onSingleTap(MotionEvent e){
        if (type.length() > 1 && type.equalsIgnoreCase("POINT")) {
            //创建 Graphic 对象
            Graphic graphic = new Graphic(mapView.toMapPoint(new Point(e.getX(),e.getY())),new SimpleMarkerSymbol(Color.RED,25,STYLE.CIRCLE));
            graphicsLayer.addGraphic(graphic);   //添加到图层中
            return true;
        }
        return false;
}
```

2）Geometry

Geometry 代表 Graphic 的几何类型（点、线、面）和空间位置信息（坐标点或坐标点串）。它有五个子类 Envelope、MultiPath、MultiPoint、Point 和 Segment，如图 4-3-3 所示。其中 MultiPath 是抽象类，它有两个子类 Polygon 和 Polyline，分别代表面和线。Segment 也是抽象类，它有一个子类 Line，代表由两个点所组成的线段。其中 Envelope、Point、MultiPoint、Line、Polygon、Polyline 这六类在开发过程中使用频率较高，因此本书着重对这六类进行说明。

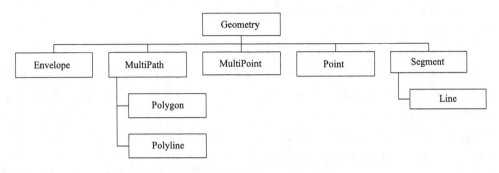

图 4-3-3　Geometry 继承关系图

（1）Envelope。Envelope 表示矩形要素，通过 Envelope 对象可获取矩形窗口的中心点、矩形的四个顶点、矩形的宽和高等。

用法如下：

Envelope env = new Envelope（112,28,113,32）；//创建矩形对象

map.setExtent(env);//设置地图显示范围

Point point= env.getCenter();//获取矩形框的中心点

（2）Point。Point 表示点对象，包括二维点对象、三维点对象。可以通过 Point 的方法设置、获取点对象的 x 或 y 坐标。

用法如下：

Point point = new Point();//创建点对象

Point.setX(114);//设置 x 坐标

Point.setY(32);//设置 y 坐标

Graphic gp = new Graphic(point, new SimpleMarkerSymbol(Color.RED, 25,STYLE. CIRCLE));

graphicsLayer.addGraphic(gp);//添加到图层中显示

（3）MultiPoint。MultiPoint 表示多点对象，MultiPoint 通常存储一系列的基础点，这些点按照一定的顺序存储。可以通过每个点的存储索引位置增加、删除或修改点对象。索引位置值自 0 开始。

用法如下：

Point point1 = new Point(114,32);//创建点对象

Point point2 = new Point(112,28);//创建点对象

MultiPoint multipoint = new MultiPoint();

multipoint.add(point1);//添加点

multipoint.add(point2);//添加点

multipoint.removePoint(1);//移除第 2 点

（4）Line。Line 表示两点之间生成的线段。Line 与 Polyline 存在一定的关系，Line 可作为 Polyline 的组成部分。

用法如下：

Line line = new Line()
line.setStart(new Point(113,32));//起始点
line.setEnd(new Point(114,28));//终止点
Polyline poly = new Polyline();
poly.addSegment(line,true);//添加线段到 Polyline 对象中

（5）Polygon。Polygon 是 MultiPath 子类，Polygon 表示的是多边形。Polygon 里的所有多边形都是闭合的环。Polygon 对象中至少存在三个点并且三点不能同时在一条直线上。

具体用法如下：

Polygon poly = new Polygon();　//创建多边形对象
poly.startPath(new Point(0,0));//添加初始点
poly.lineto(new Point(10,0));
poly.lineto(new Point(10,10));
poly.lineto(new Point(0,0));//注意多边形是闭合的，首尾点的坐标要一致。

（6）MultiPath。MultiPath 是 polygons 和 polylines 的基类，MultiPath 与 MultiPoint 很类似，只不过 MultiPoint 存储的是点的数据集，而 MultiPath 存储一条条轨迹线。

用法如下：

Point startPoint = new Point(114,28);
MultiPath path = new MultiPath();
path.startPath(startPoint);//设置起始位置
path.lineto(new Point(113,32));//添加点

3）Symbol

Symbol 是对 Graphic 对象进行符号样式设置的接口，如图 4-3-4 所示，点、线、面均有对应的 Symbol 子类。MarkerSymbol 代表点符号样式抽象类，包括 SimpleMarkerSymbol、PictureMarkerSymbol 和 TextSymbol。LineSymbol 代表线符号样式的抽象类，包括 SimpleLineSymbol。FillSymbol 代表面符号样式抽象类，包括 SimpleFillSymbol 和 PictureFillSymbol。

（1）点符号。点符号可以是简单的图形，也可以是图片或文字标注，包括 SimpleMarkerSymbol、PictureMarkreSymbol 和 TextSymbol 等三个子类。

SimpleMarkerSymbol 是针对于点状要素的 Graphic 对象进行符号设置的类，它与 PictureMarkerSymbol 类很相似，前者将点对象按照矢量点方式渲染，后者按照图片方式渲染。在使用 SimpleMarkerSymbol 类的时候，需要设置符号的属性：符号类型、尺寸大小和颜色。

用法如下：

Point point = new Point();//创建点对象
Point.setX(114);//设置 x 坐标
Point.setY(32);//设置 y 坐标
//设置点样式的颜色，大小和点类型

SimpleMarkerSymbol sms = new SimpleMarkerSymbol(Color.RED,25,STYLE.CIRCLE)
Graphic gp = new Graphic(point,sms);
graphicsLayer.addGraphic(gp);//添加到图层中显示

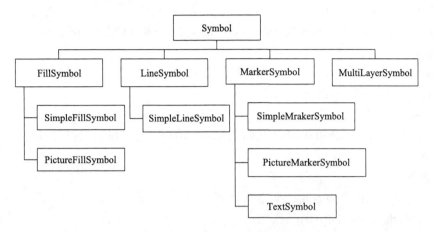

图 4-3-4　Symbol 接口类关系图

PictureMarkerSymbol 是对于点或多点要素的 Graphic 对象进行样式设置的类，可通过图片的 URL 或 Drawable 等方式来设置符号源。在使用 PictureMarker Symbol 类的时候，需要设置符号的属性：旋转角度和位置偏移量。

具体用法如下：
//创建图片样式符号

PictureMarkerSymbol pic = new PictureMarkerSymbol(getResources().getDrawable(R.drawable.icon));

Point pt = new Point(113,32);//创建一个点对象

Graphic gp = new Graphic(pt,pic);设置符号样式

graphicsLayer.addGraphic(gp);添加到图层中

TextSymbol 使用文字对要素进行渲染。可以设置其文字的大小、颜色、内容和排列方式。排列方式分为横向排列和纵向排列两种，默认为横向居中显示。

用法如下：

Point point = new Point();//创建点对象

Point.setX(114);//设置 x 坐标

Point.setY(32);//设置 y 坐标

//设置点样式的颜色，大小和文本内容

TextSymbol ts = new TextSymbol (12,"点样式",Color.RED);

Graphic gp = new Graphic(point,ts);

graphicsLayer.addGraphic(gp);//添加到图层中显示

（2）线符号。SimpleLineSymbol 是线状要素的渲染符号，可以设置符号的线型、颜色、线宽和透明度等。

具体用法如下：

Polyline polyl = new Polyline ():// 创建线对象
polyl .startPath(new Point(0,0));//添加点
polyl .lineto(new Point(10,0));
polyl .lineto(new Point(10,10));
SimpleLineSymbol sls = new SimpleLineSymbol(Color.RED,25,SimpleLineSymbol. SOLID);//线样式对象
sfs.setAlpha(50);//设置透明度
Graphic gp = new Graphic(polyl , sls);
graphicsLayer.addGraphic(gp);

（3）面符号。SimpleFillSymbol 是面状要素的渲染符号，可以设置符号的填充颜色和透明度、边界的样式等。

用法如下：
Polygon poly = new Polygon():// 创建多边形对象
poly.startPath(new Point(0,0));//添加点
poly.lineto(new Point(10,0));
poly.lineto(new Point(10,10));
poly.lineto(new Point(0,0));
SimpleFillSymbol sfs = new SimpleFillSymbol(Color.RED);//面符号对象
sfs.setAlpha(50);//设置透明度
Graphic gp = new Graphic(poly,sfs);
graphicsLayer.addGraphic(gp);

（4）Renderer。Renderer 接口主要用于 Graphic 对象的样式渲染，适用于 GraphicsLayer 的图层符号配置渲染需求。ArcGIS Runtime SDK for Android API 中提供了 Renderer 接口，有三个实现类：SimpleRenderer、UniqueValueRenderer 和 ClassBreaksRenderer，其关系图如图 4-3-5 所示。

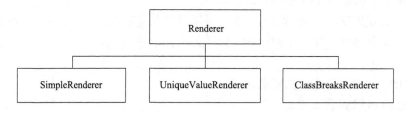

图 4-3-5　Renderer 关系图

SimpleRenderer 是最简单的一个渲染类，通过这个渲染类可以将图层要素按照同一个符号样式进行渲染。

用法如下：
SimpleRenderer renderer = new SimpleRenderer(new SimpleMarkerSymbol(Color.RED,25,STYLE.CIRCLE));
graphicsLayer.setRenderer(renderer);

UniqueValueRenderer 渲染类，是将某字段的每个唯一值配置一个对应的符号，从而完成图层渲染。例如，针对一个面状要素图层，根据面状要素属性字段"类型"的值进行渲染，类型值为"住宅楼"的符号为黄色，"工厂"的为紫色，"商业区"的为红色，也可以对多个字段进行联合唯一值渲染，最多可联合三个字段进行渲染。

用法如下：
UniqueValueRenderer uvr = new UniqueValueRenderer();
uvr.setField("TYPE");
UniqueValue uv1 = new UniqueValue();
uv.setValue(new String[]{"Residential"});
uv. setSymbol(new SimpleFillSymbol(Color.argb(128, 255, 100, 0)));
UniqueValue uv2 = new UniqueValue();
uv.setValue(new String[]{"Industrial"});
uv. setSymbol(new SimpleFillSymbol(Color.argb(128, 255, 200, 0)));
UniqueValue uv3 = new UniqueValue();
uv.setValue(new String[]{"Commercial"});
uv. setSymbol(new SimpleFillSymbol(Color.argb(128, 255, 150, 0)));
uvr.addUniqueValue(uv1);
uvr.addUniqueValue(uv2);
uvr.addUniqueValue(uv3);
graphicsLayer.setRenderer(uvr);

ClassBreaksRenderer 是分段渲染类，可以按 GraphicsLayer 图层中的某一属性字段进行分段渲染。

用法如下：
  ClassBreaksRenderer renderer = new ClassBreaksRenderer();
renderer.setMinValue(0.0);//设置最小值
renderer.setField("POP07_SQMI");//分段依据的字段
ClassBreak cb1 = new ClassBreak();//定义第一段的范围和渲染样式
cb1.setClassMaxValue(25);
cb1.setSymbol(new SimpleFillSymbol(Color.argb(128, 56, 168, 0)));
cb1.setLabel("First class");
ClassBreak cb2 = new ClassBreak();//定义第二段的范围和渲染样式
cb2.setClassMaxValue(75);
cb2.setSymbol(new SimpleFillSymbol(Color.argb(128, 139, 209, 0)));
cb2.setLabel("Second class");
ClassBreak cb3 = new ClassBreak();//定义第三段的范围和渲染样式
cb5.setClassMaxValue(Double.MAX_VALUE);
cb5.setSymbol(new SimpleFillSymbol(Color.argb(128, 255, 0, 0)));
renderer.addClassBreak(cb1);
renderer.addClassBreak(cb2);

renderer.addClassBreak(cb3);
graphicsLayer.setRenderer(renderer);//为图层设置渲染

通过上面的代码可以发现分段渲染需要以下几个步骤：

（1）创建 ClassBreaksRenderer 对象。
（2）设置渲染依据的字段和属性最小值。
（3）创建多个分段渲染对象（ClassBreak），设置该段的属性最大值和样式。
（4）将分段渲染对象添加到 ClassBreaksRenderer 对象中。
（5）为图层设置 Renderer 对象。

交互绘制操作示例图如图 4-3-6 所示。

图 4-3-6　交互绘制操作示例图

自主练习：新建一个工程，添加图层，然后添加监听器，实现在地图窗口上交互绘制点、线、面图形的功能。

实验 4-4　数据查询与检索

数据查询与检索是 GIS 空间分析的基础内容之一。移动客户端的屏幕上，通过手势操作，可以点选查询地图数据，也可以通过属性条件查询地图数据。

（1）实验目的：本实验的主要目的是了解并掌握两种常用的查询检索任务，即 Identify Task 和 QueryTask。

（2）相关实验：实验 4-3 数据显示与浏览。
（3）实验数据：本教材系列实验数据。
（4）实验环境：Eclipse4.2、ArcGIS for Android 插件、JAVA 编程语言。
（5）实验内容：本次实习主要掌握两种常用的查询检索方法，即 IdentifyTask 和 QueryTask。IdentifyTask 用于识别图层中的要素，而 QueryTask 则用于查询图层中的要素。

1. IdentifyTask 功能简介

IdentifyTask 是针对于地图服务中的多个图层中地图要素的识别，返回的结果是 IdentifyResult[]数组。它的作用就是当我们通过手指点击地图时识别获取地图上的要素信息。在识别操作前，必须为 IdentifyTask 设置好一组 IdentifyParameters 参数信息，在参数 IdentifyParameters 对象中设置相应的识别条件。IdentifyParameters 常用接口如表 4-4-1 所示。

表 4-4-1 IdentifyParameters 常用接口介绍

序号	接口	说明
1	setDPI(int dpi)	设置 map 的分辨率值
2	setGeometry(Geometry geometry)	设置空间几何对象
3	setLayerMode(int layerMode)	设置查询方式：ALL_LAYERS、VISIBLE_LAYERS、TOP_MOST_LAYER
4	setLayers(int[] layers)	设置识别的图层数组
5	setMapExtent(Envelope extent)	设置当前地图的范围
6	setMapHeight(int height)	设置地图的高
7	setMapWidth(int width)	设置地图的宽
8	setReturnGeometry(boolean returnGeometry)	指定是否返回几何对象
9	setSpatialReference(SpatialReference spatialReference)	设置空间参考
10	setTolerance(int tolerance)	设置识别的容差值

IdentifyTask 提供了三种模式，分别为：
（1）ALL_LAYERS，该模式表示在识别时检索所有图层的要素。
（2）VISIBLE_LAYERS，该模式表示在识别时只检索可见图层的要素。
（3）TOP_MOST_LAYER，该模式表示在识别时只检索最顶图层的要素。
执行识别任务需要以下几个步骤：
（1）创建 IdentifyTask 对象，设置需要访问的地图服务。
（2）创建参数对象 IdentifyParameters。
（3）定义 MyIdentifyTask 类并继承 AsyncTask。
（4）在 MyIdentifyTask 的 doInBackground()方法中执 IdentifyTask 的 execute()。
（5）处理查询结果，将查询结果封装在 IdentifyResult[]中，通过遍历 FeatureResult[]来获取其中每个图层中的空间要素。

2. Identify 实例练习

设计一个 Identify 的工程，实现当点击地图上某城市的地理位置时，弹窗显示该城市的属性信息。主要包括以下内容：

(1) 生成一个 IdentifyTask 实例对象，并且添加地图服务的 URL 路径。
(2) 获取 IdentifyParameters 对象实例。
(3) 执行 IdentifyTask 对象的 execute()方法，返回识别结果集。
主要参考代码如下：

```java
//实例化对象，并且给实现初始化相应的值
params = new IdentifyParameters();//识别任务所需参数对象
params.setTolerance(20);//设置容差
params.setDPI(98);//设置地图分辨率
params.setLayers(new int[] { 4 });//设置要识别的图层数组
params.setLayerMode(IdentifyParameters.ALL_LAYERS);//设置识别模式

//给 map 添加一个点击地图的事件监听
mMapView.setOnSingleTapListener(new OnSingleTapListener() {
    private static final long serialVersionUID = 1L;

    @Override
    public void onSingleTap(final float x, final float y) {
        if (!mMapView.isLoaded()) {
            return;
        }
        //根据点击位置添加识别参数
        Point identifyPoint = mMapView.toMapPoint(x, y);
        params.setGeometry(identifyPoint);
        params.setSpatialReference(mMapView.getSpatialReference());
        params.setMapHeight(mMapView.getHeight());
        params.setMapWidth(mMapView.getWidth());
        params.setReturnGeometry(false);

        // 增加面积范围来确定参数
        Envelope env = new Envelope();
        mMapView.getExtent().queryEnvelope(env);
        params.setMapExtent(env);

        //我们自己扩展的异步类
        MyIdentifyTask mTask = new MyIdentifyTask(identifyPoint);
        mTask.execute(params);//执行异步操作并传递所需的参数
    }
});
}
```

上面的代码主要生成了 IdentifyParameters 实例对象，并且给 map 添加了一个点击地图事件监听。

在监听器中对 IdentifyParameters 实例初始化，之后执行一个异步请求操作。

```java
private class MyIdentifyTask extends
        AsyncTask<IdentifyParameters, Void, IdentifyResult[]> {
    IdentifyTask task = new IdentifyTask(Identify.this.getResources()
            .getString(R.string.identify_task_url_for_avghouseholdsize));
    IdentifyResult[] M_Result;
    Point mAnchor;
    MyIdentifyTask(Point anchorPoint) {
        mAnchor = anchorPoint;
    }

    @Override
    protected IdentifyResult[] doInBackground(IdentifyParameters... params) {
        // 检查是否识别参数
        if (params != null && params.length > 0) {
            IdentifyParameters mParams = params[0];
            try {
                //获取要素数据
                M_Result = task.execute(mParams);
            } catch (Exception e) {
                e.printStackTrace();
            }
        }
        return M_Result;
    }

    @Override
    protected void onPreExecute() {
        // 创建对话框，同时处理 UI 线程
        dialog = ProgressDialog.show(Identify.this, "Identify Task",
                "Identify query ...");
    }

    @Override
    protected void onPostExecute(IdentifyResult[] results) {
        // 关闭对话框
        if (dialog.isShowing()) {
            dialog.dismiss();
        }
```

```
            ArrayList<IdentifyResult> resultList = new ArrayList<IdentifyResult>();
            IdentifyResult result_1;
            for (int index = 0; index < results.length; index++) {
                result_1 = results[index];
                String displayFieldName = result_1.getDisplayFieldName();
                Map<String, Object> attr = result_1.getAttributes();
                for (String key : attr.keySet()) {
                    if (key.equalsIgnoreCase(displayFieldName)) {
                        resultList.add(result_1);
                    }
                }
            }
            Callout callout = mMapView.getCallout();
callout.setContent(createIdentifyContent(resultList));
            callout.show(mAnchor);
        }
```

3. QueryTask 功能简介

通过 QueryTask 可以对图层进行属性查询、空间查询及属性与空间联合查询。QueryTask 查询任务只是针对地图服务中的一个图层进行查询。在执行 QueryTask 任务前需要构建一个包含了查询条件的 Query 参数对象。Query 的常用接口如表 4-4-2 所示。

表 4-4-2 Query 常用接口

序号	接口	说明
1	setGeometry(Geometry geometry)	设置空间几何对象
2	setInSpatialReference(SpatialReference inSR)	设置输入的空间参考
3	setObjectIds(int[] objectIds)	设置要查询要素 ObjectID 数组
4	setOutFields(String[] outFields)	设置输出字段的数组
5	setOutSpatialReference(SpatialReference outSR)	设置输出的空间参考
6	setReturnGeometry(boolean returnGeometry)	设置是否返回几何对象
7	setReturnIdsOnly(boolean returnIdsOnly)	设置是否只返回 ObjectID 字段
8	setSpatialRelationship(SpatialRelationship spatialRelationship)	设置查询的空间关系条件
9	setWhere(String where)	设置查询的属性条件

QueryTask 的使用步骤如下：

（1）创建 Query 参数对象，设置需要访问的地图服务图层。

（2）为参数对象设定查询条件，创建 QueryParameters 对象，设置查询条件、空间参考、返回字段、是否返回空间对象等。

（3）通过 AsyncTask 的子类来执行查询任务，也就是选择异步查询方式，这样在查询过程中就不会影响用户继续操作。

（4）处理查询结果，将查询结果封装在 FeatureResult 对象中，通过遍历 FeatureResult 来获取其中所有的 Feature 对象，然后做进一步的处理。

4. QueryTask 实例练习

新建一个名称为 AttributeQuery 的工程，通过示例代码来了解 QueryTask 的使用方法。
示例代码如下：

```java
querybt.setOnClickListener(new View.OnClickListener() {
    public void onClick(View v) {
        if (blQuery) {
            String targetLayer = targetServerURL.concat("/3");
            String[] queryParams = { targetLayer, "AVGHHSZ_CY>3.5" };
            AsyncQueryTask ayncQuery = new AsyncQueryTask();
            ayncQuery.execute(queryParams);
        } else {
            gl.removeAll();
            blQuery = true;
            querybt.setText("Average Household > 3.5");
        }
    }
});
```

代码中定义了一个按钮的点击事件监听，并在监听中执行我们自定义的异步类，在异步类中实现了查询功能。

执行代码如下：

```java
private class AsyncQueryTask extends AsyncTask<String, Void, FeatureResult> {
    @Override
    protected void onPreExecute() {
        progress = new ProgressDialog(AttributeQuery.this);
        //在未查询出结果时显示一个进度条
        progress = ProgressDialog.show(AttributeQuery.this, "",
                "Please wait....query task is executing");
    }
    @Override
    protected FeatureResult doInBackground(String... queryArray) {
        if (queryArray == null || queryArray.length <= 1)
            return null;
        //查询条件和 URL 参数
        String url = queryArray[0];
        //查询所需的参数类
        QueryParameters qParameters = new QueryParameters();
        String whereClause = queryArray[1];
```

```java
            SpatialReference sr = SpatialReference.create(102100);
            qParameters.setGeometry(new Envelope(-20147112.9593773,
                    557305.257274575, -6569564.7196889, 11753184.6153385));//设置查询
空间范围
            qParameters.setOutSpatialReference(sr);//设置查询输出的坐标系
            qParameters.setReturnGeometry(true);//是否返回空间信息
            qParameters.setWhere(whereClause);//where 条件
            QueryTask qTask = new QueryTask(url);//查询任务类
            try {
                FeatureResult results = qTask.execute(qParameters);//执行查询，返回查询结果
                return results;
            } catch (Exception e) {
                e.printStackTrace();
            }
            return null;
        }
        @Override
        protected void onPostExecute(FeatureResult results) {
            String message = "No result comes back";
            if (results != null) {
                int size = (int) results.featureCount();
                for (Object element : results) {
                    progress.incrementProgressBy(size / 100);
                    if (element instanceof Feature) {
                        Feature feature = (Feature) element;
                        //将功能转换成图形
                        Graphic graphic = new Graphic(feature.getGeometry(),
                                feature.getSymbol(), feature.getAttributes());
                        //将查询结果添加到图层上
                        graphicsLayer.addGraphic(graphic);
                    }
                }
                //更新结果信息
                message = String.valueOf(results.featureCount())
                        + " results have returned from query.";
            }
            progress.dismiss();//停止进度条
            Toast toast = Toast.makeText(AttributeQuery.this, message,
                    Toast.LENGTH_LONG);
```

```
            toast.show();
            // queryButton.setText("Clear graphics");
            boolQuery = false;
        }
    }
```
　　自主练习：本实验需要设计一个名为 Query 的工程，主程序可参考上述代码，设定查询的相关参数，并为 QureyTask 指定地图服务路径。

实验 4-5　数据采集与编辑

　　加载完地图以后，可进一步完成数据采集和编辑的操作，即绘制图形和编辑属性。在 ArcGIS For Android 的开发中，常使用 Geometry 类进行图形绘制。
　　（1）实验目的：初步了解数据编辑的原理和基本过程。
　　（2）相关实验：实验 4-4　数据查询与检索。
　　（3）实验数据：本教材系列实验数据。
　　（4）实验环境：Eclipse4.2、ArcGIS for Android 插件、JAVA 编程语言。
　　（5）实验内容：点、线、面等图形编辑和属性编辑。

1. 要素图层

　　从 ArcGIS 10 开始，地图服务增加了 "Feature Access" 接口，所有的要素编辑都通过这个接口实现。同时，在客户端中，增加了要素图层 "ArcGISFeatureLayer"。ArcGISFeatureLayer 其实就是和服务器端的要素图层遥相呼应的 GraphicsLayer。
　　例如：

```
        fLayer = new ArcGISFeatureLayer(
        "http://sampleserver3.arcgisonline.com/ArcGIS/rest/services/Petroleum/KSPetro/MapServer/1",o);
```

　　需要注意的是，上面示例中 ArcGISFeatureLayer 对应的 URL 地址是 MapServer 中的一个图层，该图层可以查询选择，但是不能对图层要素进行编辑。要想实现编辑功能，ArcGISFeatureLayer 的 URL 地址必须为 FeatureServer 中的一个图层。

2. Graphic 对象的属性编辑

　　可以通过 GraphicsLayer 的 getGraphicIDs(float x, float y,int tolerance)方法来获取要素及其要素的相关属性，但是 Graphic 中没有提供修改属性的接口。实现属性编辑有两种途径，一种是新建一个 Graphic 对象，并将新的属性值赋予它，调用 Feature Server 服务中的 applyEdits 接口实现更新；另一种是通过 GraphicsLayer 的 updateGraphic(intid,Map<String,Object> attributes) 的方法更新 Graphic 的属性。
　　下面的练习中，通过新建一个 Graphic 对象，并将新的属性值赋予这个 Graphic 对象，然后调用 ArcGISFeatureLayer 的 applyEdits 方法进行编辑提交，完成属性编辑。在 Feature Layer 中已经封装了 Feature Server 服务中的 applyEdits 接口，保存编辑到服务器只需要调用这个方法就可以了。
　　在本实验示例程序 AttributeEditor 中，当点击一个要素时，就会弹出一个编辑要素属性的对话框，通过该对话对现有属性值进行编辑修改，示例如图 4-5-1 所示。

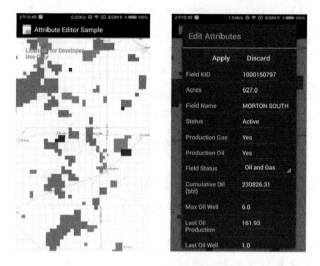

图 4-5-1　属性编辑界面示例

属性修改完毕后，点击对话框中的"Apply"按钮后，新的属性将保存到服务器。方法示例代码如下：

    Graphic newGraphic = **new** Graphic(**null**, **null**, attrs);
    …//一系列赋值
    newGraphic.setAttributeValue(featureLayer.getObjectIdField(),
listAdapter.featureSet.getGraphics()[0] .getAttributeValue(featureLayer.getObjectIdField()));
      featureLayer.applyEdits(**null**, **null**, **new** Graphic[] { newGraphic }, createEditCallbackListener(updateMapLayer));

3. Graphic 的几何图形编辑

对于 Graphic 对象，几何图形的编辑与属性编辑的思路一样，首先在客户端编辑几何对象的节点坐标，然后将更新后的几何对象提交到服务器。

在对点要素坐标修改时，通过 GraphicsLayer 的 updateGraphic(intid,Geometrygeometry)方法来实现点的空间位置更新。在对线和面的空间位置坐标修改时，通过 GeometryEngine.getNearestVertex()获得点击的线或面上一个节点，并返回一个 Proximity2DResult 对象，通过这个对象可以得到这个节点的索引（Index）位置，再通过线或面对象的 setPoint(int index,Point point)方法更新节点。

几何编辑编程的逻辑过程为：

（1）实例化对象，创建图层，并在地图中加入可绘制的 GraphicsLayer 图层。
（2）设置一个枚举类来存放命令，界面上添加按钮，当点击按钮时向系统发送命令。
（3）为地图添加监听点击事件 OnStatusChangedListener。
（4）保存修改后的数据。

在 ArcGIS Runtime SDK for Android API 中有一个可以进行空间编辑的例子（Geometry

Editor），该示例的目的是演示如何使用 ArcGIS for Android API 创建点、线、面要素，并且支持基于地图模板的三种类型（点、线和面）要素图层的编辑，包括添加、返回、保存等。

示例运行界面如图 4-5-2 所示。

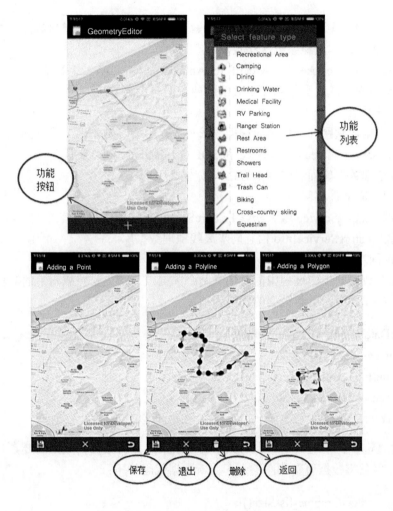

图 4-5-2　GeometryEditor 示例运行界面

图 4-5-2 中，点击底栏中的"+"图标开始添加功能，之后显示要素类型模板列表和符号，以便用户快速添加。当要添加点要素时，点击地图以定位要素位置，再次点击地图时将点移动到新的位置。当添加面或线要素时，只需轻点地图上的新位置即可添加新的顶点，然后再次点击地图上的新位置即可移动现有的顶点；需要删除现有点时，点击底栏上的垃圾桶图标即可。

实验 4-6　数据实时同步

当编辑完数据之后，下一步需要将数据上传到服务中心，以实现与服务器的同步，而数据同步则分为在线数据同步和离线数据同步两种方式。在线数据经过编辑后可以直接通过

FeatureService 服务将数据更新到图层上，进而实现编辑后数据的实时同步。离线地理数据则应根据需求与数据中心（数据库）进行同步。

（1）实验目的：本实验的主要目的是了解在线和离线两种地理数据库同步的基本思路和方法。

（2）相关实验：实验 4-4 数据查询与检索、实习 4-5 数据采集与编辑。

（3）实验数据：本教材系列实验数据。

（4）实验环境：Eclipse4.2、ArcGIS for Android 插件、JAVA 编程语言。

（5）实验内容：离线数据和在线数据两种数据同步的方法。

1. 离线数据同步

离线业务数据在联网的环境下应根据需求与数据中心（数据库）进行同步。这个交互过程需要 ArcGIS Server 的 Feature Service 和 SDE Geodatabase 配合完成。

实现离线地理数据库同步的基本思路大致为：

（1）获取 gdb 文件存储路径。

（2）根据 FeatureService 服务获取 FeatureServiceInfo 服务参数信息。

（3）根据 FeatureServiceInfo 信息同步离线地理数据库。

（4）反馈同步结果；

在离线业务数据同步的过程中需要注意两个关键点，首先是将编辑后的数据更新到数据库（即本地数据库）中，更新数据的代码为：

```
SQLiteDatabasedb=this.getWritableDatabase();
db.beginTransaction();
db.update(table,values,whereClause,whereArgs);
db.setTransactionSuccessful();
db.endTransaction();
```

之后通过查询，将本地数据库中的数据查询出来，并将数据拼成空间要素对象，再通过 FeatureService 服务将数据更新到图层上。更新数据代码如下：

```
Graphic[]gps=newGraphic[list.size()];
gps=list.toArray(gps);
featureLayerupdate.applyEdits(null,null,gps,null);
```

上述代码中 List 为 Graphic 对象的列表。

2. 在线数据同步

在线数据可以直接通过 FeatureService 服务将数据更新到数据中心进而实现数据同步。参考实验 4-5 中几何编辑（GeometryEditor）的例子，该示例中提供了在数据编辑完成之后通过点击保存并上传到服务中心的功能，实现数据实时同步。

相关示例代码为：

```
private void actionSave() {
    Graphic g;
    if (mEditMode == EditMode.POINT) {
```

```java
    //从该点开始创建一个图形
    g = mTemplateLayer.createFeatureWithTemplate(mTemplate, mPoints.get(0));
} else {
    // 对于线和面，从一个点开始创建一个 MultiPath
    MultiPath multipath;
    if (mEditMode == EditMode.POLYLINE) {
        multipath = new Polyline();
    } else if (mEditMode == EditMode.POLYGON) {
        multipath = new Polygon();
    } else {
        return;
    }
    multipath.startPath(mPoints.get(0));
    for (int i = 1; i < mPoints.size(); i++) {
        multipath.lineTo(mPoints.get(i));
    }
    //然后简化几何图形并从中创建一个图形
    Geometry geom = GeometryEngine.simplify(multipath, mMapView.getSpatialReference());
    g = mTemplateLayer.createFeatureWithTemplate(mTemplate, geom);
}
//显示进度条并在保存期间禁止操作
setProgressBarIndeterminateVisibility(true);
mEditMode = EditMode.SAVING;
updateActionBar();
//将编辑后的图形添加到图层
mTemplateLayer.applyEdits(new Graphic[] { g }, null, null, new CallbackListener<FeatureEditResult[][]>() {
    @Override
    public void onError(Throwable e) {
        Log.d(TAG, e.getMessage());
        completeSaveAction(null);
    }
    @Override
    public void onCallback(FeatureEditResult[][] results) {
        completeSaveAction(results);
    }
});
}
```

上述代码实现了将编辑之后的数据上传给服务中心，之后需要"保存"操作的结果，即上传给服务中心实现数据同步的结果反馈给用户并退出编辑模式。

示例代码为：

```
void completeSaveAction(final FeatureEditResult[][] results) {
    runOnUiThread(new Runnable() {
        @Override
        public void run() {
            if (results != null) {
                if (results[0][0].isSuccess()) {
                    String msg = GeometryEditorActivity.this.getString(R.string.saved);
                    Toast.makeText(GeometryEditorActivity.this, msg, Toast.LENGTH_SHORT). show();
                } else {
                    EditFailedDialogFragment frag = new EditFailedDialogFragment();
                    mDialogFragment = frag;
                    frag.setMessage(results[0][0].getError().getDescription());
                    frag.show(getFragmentManager(), TAG_DIALOG_FRAGMENTS);
                }
            }
            setProgressBarIndeterminateVisibility(false);
            exitEditMode();
        }
    });
}
```

进行完这些操作之后即可实现数据实时同步，并可在客户端显示编辑更新后的结果。

示例应用界面如图 4-6-1 所示。

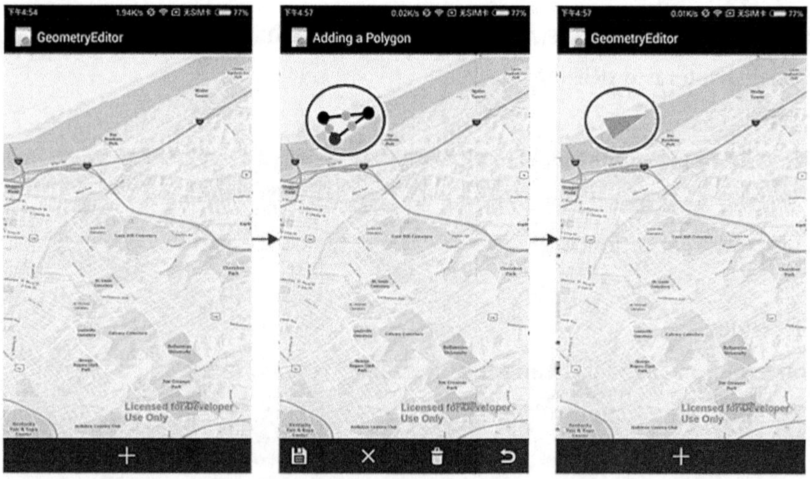

图 4-6-1　在线数据同步示例应用界面

主要参考文献

艾明耀, 胡庆武. 2017. 高级 GIS 开发教程. 武汉: 武汉大学出版社.
陈永刚. 2016. 开源 GIS 与空间数据库实战教程. 北京: 清华大学出版社.
冯增才. 2016. 地理信息系统 GIS 开发与应用. 天津: 天津大学出版社.
兰小机, 刘德儿, 魏瑞娟. 2011. 基于 ArcObjects 与 C#. NET 的 GIS 应用开发. 北京: 冶金工业出版社.
李仁杰, 张军海, 胡引翠. 2018. 地图学与 GIS 集成实验教程. 北京: 科学出版社.
刘光, 唐大仕. 2009. Web GIS 开发 ArcGIS Server 与. NET. 北京: 清华大学出版社.
柳锦宝, 张子民, 张永福, 等. 2010. 组件式 GIS 开发技术与案例教程. 北京: 清华大学出版社.
罗显刚. 2015. 网络 GIS 行业应用开发实践教程. 武汉: 中国地质大学出版社.
马林兵, 张新长, 吴苏杰, 等. 2019. Web GIS 技术原理与应用开发. 3 版. 北京: 科学出版社.
马云强, 杜婷, 毕猛, 等. 2013. 基于 DotSpatial 的轻量级 GIS 开发技术研究. 云南地理环境研究, 25(3): 39-44.
牟乃夏, 王海银, 李丹, 等. 2015. ArcGIS Engine 地理信息系统开发教程. 北京: 测绘出版社.
戚正伟, 付国庆, 蔡松露, 等. 2009. 嵌入式 GIS 开发及应用. 北京: 清华大学出版社.
芮小平, 于雪涛. 2015. 基于 C#语言的 ArcGIS Engine 开发基础与技巧. 北京: 电子工业出版社.
吴信才. 2013. 搭建式 GIS 开发. 北京: 电子工业出版社.
张丰, 杜震洪, 刘仁义. 2012. GIS 程序设计教程——基于 ArcGIS Engine 的 C#开发实例. 杭州: 浙江大学出版社.
张贵军, 陈铭. 2016. WebGIS 工程项目开发实践. 北京: 清华大学出版社.
赵建三, 邓丁杰, 唐利民, 等. 2013. 基于 DotSpatial 的组件式 GIS 开发与应用. 软件, (12): 119-121.
周林, 陈占龙, 罗显刚. 2014. 网络 GIS 开发实践入门. 武汉: 中国地质大学出版社.
Scott Davis. 2008. GIS for Web 应用开发之道. 蒋波涛译. 北京: 电子工业出版社.